THE ANTARCTIC BOOK OF COOKING AND CLEANING

// A POLAR JOURNEY //

WENDY TRUSLER AND CAROL DEVINE

PRINCIPAL PHOTOGRAPHY
SANDY NICHOLSON

HARPER
DESIGN
An Imprint of HarperCollins Publishers

For Fin, Sasha and Veronica

First published in Canada in 2013 by Vauve Press.

The Antarctic Book of Cooking and Cleaning
Copyright © 2015 by Wendy Trusler and Carol Devine

Published in 2015 by Harper Design
An Imprint of HarperCollins Publishers
195 Broadway
New York, NY 10007
Tel: (212) 207-7000
Fax: (855) 746-6023
harperdesign@harpercollins.com
www.hc.com

Distributed by HarperCollins Publishers
195 Broadway
New York, NY 10007

ISBN 978-0-06-239503-0

Library of Congress Control Number: 2014951625

Editor: David Young
Recipe editor: Donna Bartolini
Book design by The Office of Gilbert Li
Principal photography by Sandy Nicholson

Printed in China
First Printing, 2015

www.theantarcticbookofcookingandcleaning.com

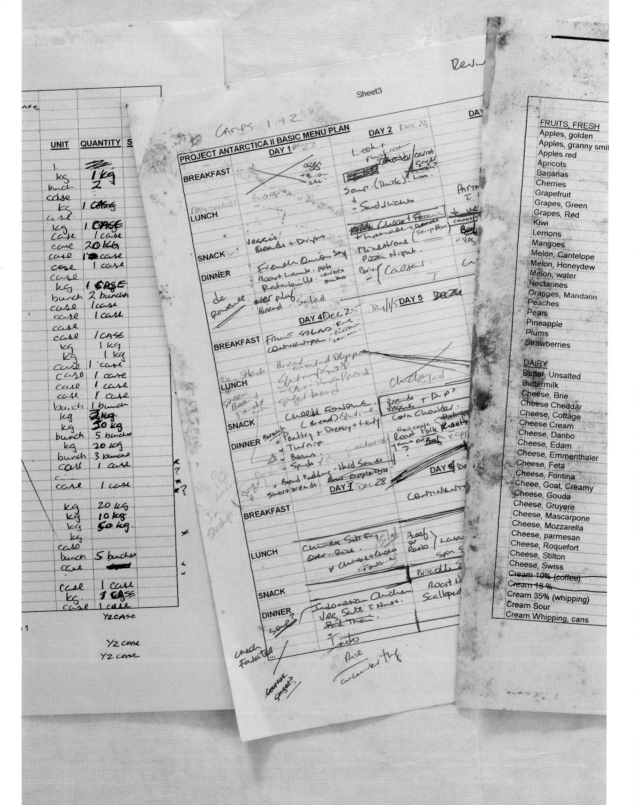

Wendy's provisioning notes, 1995

PREFACE

Carol Devine & Wendy Trusler

———

The first thing that comes to mind about Antarctica is not likely food. But if you are going to spend any time there, it should be the second.

In 1996, we led several volunteer groups to Bellingshausen, a Russian research station in Antarctica, for an environmental project organized in collaboration with the Russian Antarctic Expedition. A total of fifty-four people from five countries paid to pick up twenty-eight years of garbage during their holiday on a continent uniquely devoted to peace and science.

The Antarctic Book of Cooking and Cleaning is a journey through that austral summer—the story of a Russian-Canadian cleanup project on a small island 120 kilometres off the Antarctic Coast (62°02′s 58°21′w). It is also a look at the challenges of cooking in a makeshift kitchen during the long, white nights at the bottom of the world.

The book unfolds in the style of Antarctic publications such as Sir Ernest Shackleton's handmade *Aurora Australis*—through provision lists, menu plans, journals and letters. Woven throughout are historic and contemporary images, food and science notes as well as vignettes from Antarctica.

Early explorers and scientists endured unimaginable conditions, surviving on penguin meat and even dog paw stew. *The Antarctic Book of Cooking and Cleaning* is an ode to them, to contemporary explorers and scientists and to the people we met who captured our hearts.

It has been a curious and beguiling exercise for us to reflect on our experiences in a male-dominated fish bowl.

Whenever adventure beckons an open mind and a full stomach are necessities.

MAGNETISM

"Yet as those noble peaks faded away in the mist, I could scarce express feelings of sadness to leave the land that has rained on us its bounty and been our salvation."

—Frank Hurley, 1885–1962

Frank Hurley photographing under the bows of the *Endurance*, 1915.
Glass Paget plate photo transparency. Before abandoning the ship, Shackleton and Hurley chose 120 glass plates to keep, including this rare colour one. They smashed 400 plates; Shackleton feared Hurley would endanger himself by even thinking of returning for them.

INTRODUCTION
Carol

I always wanted to go to Antarctica without knowing why. I grew up in the subarctic and often went to school with icicles in my hair. I used to press one hand on the top of the globe and with the other spin it so quickly that the water and continents blurred together. What was that odd-shaped white continent down there on the bottom? Ptolemy mapped Antarctica in antiquity before anyone even knew it existed. There was only intuition that a continent must exist at the bottom of the world counterbalancing the land masses at the top. Humans have inhabited Antarctica only in the last 100 years.

Antarctica occupies a tenth of the earth's surface. Scientists tell us that the icy continent most resembles the moon of Jupiter: Europa—also thought to hold a liquid ocean beneath its ice. None of us are going to Europa, and only a tiny sliver of the human population will ever be lucky enough to visit Antarctica. I was one of the fortunate few. My first moments on the continent felt as if I had reached both the moon and a kind of paradise amidst those baby blue icebergs. It was a place where each footprint felt momentous and each interaction vividly important.

Antarctica holds the majority of the world's ice and fresh water—it is critically important to the planet and is a mirror reflecting what we humans mean for the Earth. Nature's laboratory in the south reveals things like a record of the earth's climate held in its 420,000-year-old ice core. What happens in Antarctica doesn't stay in Antarctica and vice versa. The Antarctic Peninsula and the surrounding oceans have warmed faster than anywhere else in the Southern Hemisphere. It is evident, in the first moments one spends here, that the continent bears devastating marks of human activity.

As a nascent humanitarian, I wanted to do something previously untried—to set up a project that would give Antarctic visitors an opportunity to help conserve the environment in that wild, remote place.

I approached Sam Blyth who ran a company that took adventurers to the Antarctic and together we created the VIEW Foundation—Volunteer International Environmental Work. Pat Shaw, Vice-President of the polar travel side of Sam's company, Marine Expeditions International (MEI), suggested a VIEW project in Antarctica and helped lay the groundwork.

For starters, we needed a partner in Antarctica. I wrote letters to the appropriate authorities in the U.K., Russia, Poland and Australia asking if they would be interested in having international volunteers assist with conservation efforts at their Antarctic stations.

On June 16, 1994, I received a handwritten letter from the Polish Academy of Sciences in Warsaw. At the Henryk Arctowski station in the South Shetland Islands, the scientists were already starting to consolidate debris that had accumulated at its base since opening in 1977. This first collaborative environmental initiative would become Project Antarctica.

MEN WANTED

for hazardous journey, small wages, bitter cold, long months
of complete darkness, constant danger, safe return doubtful,
honor and recognition in case of success.

Ernest Shackleton 4 Burlington st.

Fabled ad in London newspaper circa 1914 for the Imperial Trans-Antarctic Expedition, 1914–1916

TRAVEL ADVISORY: ANTARCTICA;
SOUTH POLE CLEANUP

The New York Times Archives

January 29, 1995
Sightseeing is, of course, allowed, but travelers who sign on for a 12-day trip to Antarctica
will spend much of their time cleaning up a research station.

Organized by VIEW, a Canadian conservation group, the trip, from March 10 to 21,
entails four days of not-too-hard labor, like putting light debris into garbage bags, at the
Polish Research Station.

From Buenos Aires, travelers fly to Ushuaia, Tierra del Fuego, then board the Akademik
Sergey Vavilov, a 79-passenger Russian research vessel, for Antarctica.

The trip—$1,800 a person plus $1,102 round-trip air fare from New York to Buenos Aires—
includes lectures and a cocktail party. Information: (416) 964-1914.

Next we needed volunteers able and willing to pick up garbage and cover their own travel expenses. I faxed a press release to major newspapers—*The New York Times*, *The Toronto Sun*, *The Boston Globe*, the *Arizona Daily Star*—and others picked it up. The phone rang and we soon had our numbers: fifty-seven people, including teachers, retirees, a judge, a music industry executive, two writers, a Dutch travel agent and an amateur cartoonist.

Problems and bad weather beleaguered our pilot expedition. A delayed departure from Argentina and landing late at Arctowski shortened our project, but we nonetheless managed to help collect and remove 16 tonnes of debris and scientific equipment back to Poland.

This initial success motivated me to set up another expedition, Project Antarctica II, this time over one summer and with volunteers living at a base. I was nervous about the logistics and recruitment for a bigger cleanup but wanted to persevere and see how the concept could develop.

The Russian Antarctic Expedition (RAE) offered to host VIEW volunteers. I flew to Saint Petersburg to negotiate the terms of our collaboration. I met with Dr. Valery Lukin, the director, and Victor Pomelov, the environment manager, in the map-filled room at the Arctic and Antarctic Research Institute. I'm sure they expected someone older and more experienced, but they agreed to the cleanup at Bellingshausen. The RAE was likely keen to generate some new money for their Antarctic program by charging a reasonable fee to host our volunteers. Lena Nikolaeva, their colleague, translated and I suspected helped close the deal. She winked at me as I left.

She would be seconded to act as liaison officer. This delighted me; Lena had fantastic energy and significant Antarctic experience.

One condition was that we had to bring our own cook. I was daunted. I had no idea where I was going to find someone to spend three months hustling meals at a dilapidated Russian base in the Antarctic. A friend recommended Wendy Trusler, renowned for her cooking at tree-planting camps across the Canadian north.

I called Wendy immediately and asked if she wanted to interview for the job. She said yes and I soon met her: there she was looking (very unconsciously) like Amelia Earhart.

A few days later I savoured her honey oatmeal bread and held the beautiful hand-carved wood burls she showed at her solo art exhibition. I offered Wendy the job, scoring not only a cook but also a respected visual artist. We were a go...almost. I needed a program manager and hired a Scot, Antarctic veteran Sean Stephen. I also hired American biologist Dr. John Croom to round out our crew. And we had the wonderful Lena as a cultural bridge. Each had an essential role, but the egg in the cake was Wendy.

Volunteers were signing up; our numbers for Project Antarctica II were high enough to be financially viable: fifty-four. We were all conscious of the scrutiny we could receive by having tourists stay at research stations. It was my job to ensure that we didn't interrupt scientific work and that we left the physical environment looking better than we found it.

Over the next three months a tourist ship would do scheduled drop-offs of small groups of volunteers at Bellingshausen for four-to-five day stints of garbage collection. The VIEW Foundation team was constant: Wendy, Lena and Sean, and for the earlier period, John and I.

Project Antarctica, VIEW Foundation
pilot cleanup at Henryk Arctowski (Poland) station,
Carol Devine in centre, 1995

Arctowski station, 1995

Bellingshausen is part of the South Shetland Islands group off the Antarctic Peninsula. Built in 1968, Bellingshausen station became a major fuel depot for the Soviet Antarctic fishing fleet. When we arrived, the base was a cluster of paint-chipped prefab buildings set amid a scattering of derelict machinery.

Several nations have research stations on this 90-percent glaciated island, including Russia, Chile, Uruguay, China, Poland, Argentina, Peru, Korea and Brazil. Evidence shows glaciers there and elsewhere are rapidly retreating. Every scientific station in Antarctica faced the problem of garbage—and they were all under pressure to comply with the Madrid Protocol on Environmental Protection (1991) and meet its stringent rules, including waste disposal, phasing out open incineration, and sewage treatment for bases with more than 30 summer staff.

We discussed garbage with the Russians and volunteers while picking it up by hand in sleet and intermittent sunshine. How do you start to clean up some 28 years worth of accumulated rubbish and encourage long-term commitment to a cleanup? We knew environmental protection depended on politics, leadership and commitment.

Governments and private businesses have designs on the riches of Antarctica—gold, uranium, gas and oil reserves and more. Despite the stated global commitment to devote Antarctica to peace and science, the military-industrial complex lurks around every corner. The Chilean air force runs the base next door to Bellingshausen. The current U.S. Antarctic logistics contractor is Raytheon, a major producer of advanced weapons systems.

The Russians told us they needed additional money and political support to do a significant environmental cleanup. They certainly weren't as hyper-conscious about littering or stepping on moss as we were. They were openly burning garbage when we arrived. But the Russians demonstrated a largely waste-free, simple style of living and, just as valuable, boundless hospitality. During our Bellingshausen summer they shared their home and welcomed our people with open arms.

This book is an invitation to experience our and others' passions, doubts, victories, disasters, concerns, joys, heartbreaks, discoveries, recipes, warnings and encouragement for crossing stormy passages and being (or at least trying to be) good citizens of the world. It's a call for earth stewardship. Why should future generations have to clean up our collective mess and inherit a planet depleted of biodiversity and resources?

Food is life, food is culture. It shaped old expeditions and shaped ours, and we're going to use it to tell you this story.

CASTING OFF

AUTUMN IN CANADA 1995

The last time I sent a job offer to someone
I wrote at the top of the job description
"For an elegantly wild person who likes adventure."

SEPTEMBER 18, 1995 // SAINT PETERSBURG

CAROL   Negotiating Project Antarctica II. I met Victor Pomelov, environmental manager of the Russian Antarctic Expedition, first. Nice guy. We had a meeting and a vodka at Hotel Sovetskaya, next to the Fontanka river. I love it here. My third trip to Saint Petersburg and by chance each time I stay at or near the same hotel.

The next day: I went to the *Arctic and Antarctic Research Institute* at 38 Bering Str and also met Dr. Lukin, the director and Lena Nikolaeva who translated our meeting. She has caramel short hair, is in her early forties, friendly. Lena has a good laugh.

The office had wood-panelled walls and was bustling. There were polar maps and photos everywhere. I believed in the expedition's purpose. Beyond the humble environmental assistance we could provide alongside the Russians, we could have a meaningful cross-cultural and educational experience. They seemed willing to try.

We agreed on the terms for what we ultimately called The Joint Russian Canadian Ecological Project: how much we paid to live at their station, where in their bases we could be accommodated, that we provisioned our own food and we had our own emergency plans, and how we would help their conservation efforts. Lena's translations of my words to Dr. Lukin were longer than the sentences I had said in English.

The previous year Lena had crossed Antarctica to Novolazarevskaya (Novo) station with the Russians in a train-tractor vehicle; she showed me a photo of the tractor. That's hard-core.

She'd also been a radio operator in Dronning Maud Land with the Norwegians on a mountaineering expedition at Wohtat Massive near Novo. Lena told me she participated in a section of a four-year Novo cleanup.

Part of our agreement was to have Lena join us at Bellingshausen as our liaison with the Russians. I liked that proposal and I knew we'd get along well.

Victor took me to the "Museum of the Arctic and Antarctica" in a restored old yellow church. An impressive display of black and white photographs and Arctic paintings, a replica of an Arctic drift ice-based station and dioramas. More than one stuffed penguin.

Now the project feels very real.

OCTOBER 5, 1995 // TORONTO

I'm meeting Wendy today. Bryce said his friend had thrived on her cooking last summer. There is so much to still work out to make this so-called expedition happen. But we succeeded despite all last year, somehow.

Her résumé: "A visual artist with interests spanning painting, photography, and sculpture, Wendy Trusler is also owner/operator of her own catering company, est. 1988. Renowned for her expertise cooking in remote work sites. Her versatility means she is as comfortable catering a cocktail party in Toronto as she is preparing hearty meals for hungry tree-planters in a northern Ontario bush camp." I had a good feeling.

Later: Wendy was fit, with long auburn hair, curly. We sat in a windowless office with a photo of an Estonian icebreaker ship on the wall. She invited me to her upcoming art show *Forest Stories*. I wanted to hire her then and there. Imagine two-in-one: a cook and resident artist.

OCTOBER 18, 1995

Interviewed Sean today by telephone for the camp manager position—to take over after I leave Antarctica. Why does everyone I interview sound so casual but interested in the job? He's a Scottish mechanical engineer who loves mountaineering. Worked for the Scott Polar Research Institute and The Antarctica Project at Cuverville Island studying penguins. I was happy Sean also knew the Antarctic Treaty, no-trace camping and the sanctity of Sites of Special Scientific Interest, as there are a few near Bellingshausen. I worried he was too skeptical about our cleanup aims.

I also called Greenpeace in New Zealand, as their *MV Greenpeace* was in Antarctica earlier this year inspecting bases. They kindly said they'd mail me the confidential report and public materials. This can help give me a picture of what we can do at Bellingshausen and the concerns.

OCTOBER 19, 1995 // TORONTO, 401 RICHMOND

Opening of Wendy's show. The gallery smells like a forest. Wendy's burls are smooth, carved wood, those big knobs that grow on trees. I told her the job was hers if she wanted it.

The last time I sent a job offer to someone I wrote at the top of the job description "For an elegantly wild person who likes adventure." Wendy said yes.

OCTOBER 20, 1995

To do:

· send Wendy a contract
· send Sean a contract
· call Barb about Nike donation: expedition jackets, hats
· fax Lena. Bring walkie-talkies? Number of bedrooms for volunteers?
· gear from Canadian government—do they have sub-zero sleeping bags, tents, stove?
· information packages for volunteers
· make appetizer for Heidi's party Friday

NOVEMBER 2, 1995 // TORONTO

Provisioning day with Wendy.

She took me to shops on Chinatown's Spadina Avenue that I have never been in before. I carried the money and Wendy chose small and large cooking utensils of shapes and purposes unrecognizable: cutlery, plates, mixing bowls. She was meticulous and thrifty, nothing too large, nothing too expensive, nothing impractical, she assured me. How the hell were we going to get these things to Antarctica?

We knew what we'd face couldn't be anything as serious or tough as the challenges endured by old-school Antarctic explorers like Adrien de Gerlache or Ernest Shackleton, but despite understanding scurvy and having the benefit of modern communication, technology and flights to the Antarctic, the variables were still many. It was a make-it-up-as-you-go enterprise based on a shared belief it was worth it. Wendy invited me over afterwards. We had a glass of wine at her place, a warm and narrow house with antique wooden skis, old hat boxes and cookbooks about. For the first time we relaxed. Looked through Antarctic books donated by a geography teacher, a former VIEW volunteer. Got excited about the upcoming trip.

NOVEMBER 4, 1995

Bonus: Dr. John Croom will join us for part of the season as resident scientist. He sounds accomplished and experienced, good slow southern U.S. drawl and he's studied at Bellings-hausen before. Talkative. Was research program biologist at U.S. Palmer Station in 1968–69 and then went to what was Soviet Bellingshausen station as exchange scientist in 1970. I asked him, "So, what is there to clean up?" He laughed and said, "Well, likely things from the seventies." John is an associate professor at Emory University, Georgia, in the human and natural ecology program. The Croom Glacier at 70°18′ 62°25′ w on the east coast of Palmer Land was named after him. John said he plans to measure water quality and sample bacteria during his stay at Bellingshausen. Later he'll investigate if the bacteria can be used to break down petroleum products in the soil.

NOVEMBER 7, 1995

I laugh and do new things with K such as play basketball. But he mentioned children. Not a good idea. We've barely started dating.

A mix of Americans, Canadians and a few other nationalities are coming to Bellingshausen. Maybe the average age is fifty-two. A lot of seniors go to the Antarctic, those with the income, but I think our project attracts the healthy, fit and adventurous. I hope so. It's also not a crazy price to get both a tour and a unique volunteer opportunity. There has to be something great about most people who pay for an unusual work experience.

It's hard to answer volunteers' questions on where they are going to sleep or what work they'll do. I say what I can and make up the rest using the disclaimer, "For example, you might..." The majority don't expect detail. Really I should go to Bellingshausen first but there is no time or budget. I hope there are enough rooms and beds for everyone.

The constant preoccupation is what will be the cleanup work. I asked Lena a few times when I was in Saint Petersburg. She said, "Don't worry. There is a lot to do." We have to wing it. At Arctowski earlier this year it was a mix of heavy labour and light.

NOVEMBER 11, 1995
It's good our office is next to Marine Expeditions Inc. They lend us services like booking the ship cabins, flights to Ushuaia and hotels in Buenos Aires for the volunteers. Their graphic designer made us an expedition T-shirt and badge.

NOVEMBER 20, 1995
I called New York. In Saint Petersburg Lena told me about Bruno "Penguin" Zehnder, a photographer and ecologist who recently lived at Russia's Antarctic Mirny station and is helping with a cleanup there. I like learning of other cleanup initiatives. He sounds like a character. Bruno wants to meet us en route to Argentina and says it's excellent we're working with Bellingshausen; every bit counts. He said it is important to work collaboratively with the Russians.

How it's going to work: Marine Expeditions offers us a rotating schedule on the *Akademik Boris Petrov* and the *Professor Multanovskiy* to drop off and pick up passengers on a five-day schedule. MEI will accept two 50 gallon barrels of non-toxic, non-hazardous debris from the Bellingshausen cleanup on the ship with each volunteer group returning to Ushuaia.

Greenpeace report arrived. In handwriting in red pen across the top: Preliminary Report, January 1995 CONFIDENTIAL: NOT FOR EXTERNAL USE.

The press release on the base inspections is headlined, *Greenpeace Expedition Finds Little Evidence of Effective Implementation of Environmental Agreement in Antarctica.* They noted effort and money were put into the logistics of maintaining a presence in the Antarctic rather than into science and reported "a marked increase in military activity in this ostensibly de-militarized region of the world."

It's a helpful internal report but I was surprised read to a reference to Dostoevsky in it.

Russia is planning to ratify the Madrid Protocol. Hopefully all Antarctic nation's environmental practices will improve. Surely the bases are a reflection of the government's current interests.

Antarctica coming up fast. Excited to return. So much to do, K frustrated I am so busy. Promised to go skating with him tomorrow night.

Fat snowflakes like bits of tissue paper are whirling around outside my window.

ALL THESE YEARS LATER...
Wendy

———

I see the opening for my installation Forest Stories as a watershed moment. I had made the break from cooking for tree-planting camps the previous spring to join a team paddling across Canada to raise awareness for AIDS. And when that venture didn't pan out—too proud to head back to the bush—I spent my first summer in nine, away from the Canadian boreal forest, creating a body of work to honour it. Fifty bowls hand-carved from wood burls salvaged from slash piles up north.

The opening was packed. Family and friends, many of them tree-planting alumni, gathered in the woodland I recreated in the gallery. Carol let me know I got the job as soon as she arrived and a "You're going where?" buzz added to the night. I felt sorry word reached my parents before I had a chance to tell them. Still smarting from the cancelled canoe trip, I had kept this project under wraps since I applied in September. I shared the news with my boyfriend and thanked the friend whose exaggerated endorsement, "You are the only person who can get food to Antarctica," had given me confidence to pursue the position.

From the outset the project struck a chord. My work in the North had a similar enviro bent and I enjoyed the challenge of making things run smoothly in seemingly unworkable conditions. Having few resources to draw on brings out the best in some people; that's my tribe. But for all its romantic appeal I accepted the job because devoting the growing season to making art hadn't been practical: I needed the money.

Carol and I worked on logistics immediately. By late October she had me compiling wish lists. What would I need for the kitchen, dining room or common room? What special things would I want to do? Her attention to detail was extraordinary; I could tell she cared deeply about what we would be doing and the people who would be participating in the project, whoever they were.

Apart from a journal to navigate occasional rough spots in my twenties I was never much of a diarist, but I am a list-maker and keeper of things (at least until I understand what they want to be). From calendars, letters, and lists I've kept all these years I can trace my mindset leading up to the project. In a way they tell me more than a diary could.

In early November the report arrived from a five-year post-op brain scan I had that summer—a milestone to remind me how resilient and vulnerable we are. No tumour recurrence, just bit of scar tissue responsible for occasional nocturnal short-circuits—no real cause for worry. I was good to go.

On November 22nd our first purchase order was to be faxed to our supplier in Ushuaia, Argentina. My cantina Spanish ("Dos cervezas, por favor") was put to the test.

In all the busyness of goodbye it's easy to neglect making time to say it. As November drew to a close P organized a surprise going-away party. We were in a newish relationship—his first since a separation and my first in four long years getting my footing after a love that didn't hold. In the weeks after I accepted the job we developed a confusing pattern of twelve-hour breakups. I assumed he was dealing with emotional baggage and let myself think he loved me more than anyone else had.

Filling a restaurant with well-wishers was a lovely gesture, but I felt silly about all the fuss, given I was only going to work. Mostly I was touched he thought of it and took it as a sign that he was becoming more comfortable with my migratory patterns.

You can fit a lot more than skis in a ski bag. In addition to my knives, aprons, food processor, a few key cookbooks and music I also packed my stuffed bears, a quilt I made from my old T-shirts, inspiring non-fiction, Russian and Spanish phrase books, art supplies and telemark ski equipment. I make home wherever I am.

Art supplies were the real luxury. For all the freedom it brought me, the drawback of cooking for a third of the year was shutting down my art practice. With fewer mouths to feed than in tree-planting camps I was optimistic I'd have time for art—perhaps even for the project I was leaving midstream. As I folded my studio, I got the idea to collect Antarctic recipes to incorporate into the cookbook/painting I was making with recipes from my tree-planting days. Carol gave me a small recipe file to get me started and asked me to write a teaser about the cookbook to include in my VIEW team bio.

Not one of my traditional mediums, writing was a bit of a leap, especially since I rarely found time to write letters while working. But for some reason one of the last things I did to prepare for the journey was arrange to write to CBC Radio host Peter Gzowski for his Letters to Morningside broadcasts. I also bought a journal. Soon we were on our way.

Books:
A Schoolteacher in Old Alaska: The Story of Hannah Breece, Jane Jacobs ed.
Roughing it in the Bush, Susanna Moodie
The Miracle of Language, Richard Lederer
Northern Bounty: A Celebration of Canadian Cuisine, Jo Marie Powers and Anita Stewart eds.
The Cook's Handbook, Prue Leith
The Enchanted Broccoli Forest, and The Moosewood Cookbook, Mollie Katzen
Potager: Fresh Cooking from the French Kitchen, Georgeanne Brennan
South to the Pole, L.B. Quartermain
Under Scott's Command: Lashly's Antarctic Diaries, Commander A.R. Ellis ed.
Scott's Last Expedition: The Journals, G. C. Scott

"The first three weeks of November have gone by with such
a rush that I have neglected my diary and can only patch it up
from memory."

—Robert Falcon Scott, *Scott's Last Expedition, Volume 1,* November 25, 1910

Frank Hurley: The Bosun (John Vincent) of the *Endurance* mending a net.
Glass Paget plate phototransparancy, 1915

MIRRORS

Carol

———

During the spring of 2011, I visited the Scott Polar Research Institute (SPRI) Library and Archives at the University of Cambridge. Archives are time travel. I went with the question, "What were those early explorers thinking?"

This archive was a small room with four desks and two private doors: one to the collection and one to the picture library. I sent my list of requests for documents. I waited.

With white gloved-fingers I opened Robert Falcon Scott's 1910 notebook; its pages were empty, but for three words crossed out in pen: ~~Limes from Melbourne~~.

I found tidy notes in pencil—the winter lectures of the Terra Nova Expedition on birds and meteorology, including Antarctic rainbows.

In a simple black journal was Scott's chart of time passing and list of food to be consumed en route to the South Pole. One note in pencil on grey-lined paper was, "1. Large proportions of butter." In one calculation Scott appeared to have foreshadowed missing the return ship. Scott and his party reached the South Pole a month after Roald Amundsen's Norwegian expedition won the race on December 14, 1911.

On Scott and party's trek back to the ship they died, only eighteen km from a food cache.

One document was a letter with the loopy, confident handwriting of Peggy Pergrine. She asked Shackleton if she could join his 1914 Imperial Trans-Antarctic expedition. Shackleton refused and it was remarkable any of the men who did returned. Their ship, the *Endurance,* got crushed in pack ice, stranding twenty-eight men in no-man's land.

These days, who doesn't enjoy a survival story. Australian explorer and artist Frank Hurley's hopeful, poetic entries about science and food in his journal from Shackleton's expedition gripped me such as, "In the stomach of the leopard were found some 50 pre-digested fish, in excellent condition, their stomachs in turn, crammed full with amphipods." (March 30, 1916). I raced to read about the starving party's rescue from desolate Elephant Island. This island is near King George Island where we lived in relative luxury 80 years later.

Arctic explorer, scientist and humanitarian Fridtjof Nansen (1861–1930) inspired me most. He shared survival lessons learned from the Inuit and participated in creating the League of Nations. He designed the Nansen Passport that was issued to 450,000 stateless people and honoured by 52 governments.

I travelled further reading journals that revealed more of the writer's feelings than only details of the temperature or meals. The observations were entryways, whether they were written to be shared or not.

I discovered parallels in Wendy's and the historic journals: the beauty of our shared humanity, records of the weather and heart, humour and hardship, the shifting inside and outside world, the value of knowledge transfer and a hearty stew.

EXPECTATION

DECEMBER 6–20, 1995

Outside it's just water, sky and seabirds—
we're all waiting for the first iceberg sighting.

DECEMBER 6, 1995 // TORONTO

CD   Everyone says, "Be careful".

We have nine bags weighing 40 or 50 pounds each.

"As they go by tell me what is in them," says the airport customs officer.

"Pots, pans, utensils, books," says Wendy as the first bag goes through the X-ray machine.

"More pots, pans, utensils, books, and skis," she continues.

"This big one?" the officer asks.

"Carol's makeup," Wendy teases.

P and K took forever saying goodbye. The guys dipped us at the same time in an embarrasing tango move. The customs guy said, "Okay, time to go. Leave them alone."

WT   Voyage to el fin del mundo. Early morning pickup at my place so I'm tired before we start. No time for breakfast or dishes. Keys in the mailbox for subletters. Last minute visit next door to drop off painting. Squish into the airport limousine. Wait at airport surprisingly emotional given how weird things have been with P recently.

Commuter plane to NYC. Carol gets hit on. At airport met Bruno Zehnder—gave us slides of his cleanup at Mirny station and copies of his emperor penguin calendar to sell to the "rich tourists".

NYC switch and then Miami to Buenos Aires.

Ushuaia most spectacular, a sharp contrast to arid land we've flown over.

Note from Dave German welcoming us then taken directly to ship. Funny to run into friends from the tree-planting world down here.

DECEMBER 7, 1995 // USHUAIA

CD   Flying into Ushuaia was amazing. We did a dramatic swooping landing on a narrow airstrip next to the water. Snowy mountains behind the town.

The wooden airport terminal is charming. A local Marine Expeditions staff member, Pablo, holds up a sign for Wendy and me. He takes us straight to the ship, the *Petrov*. Ian Shaw greets us. He's working on the ship. Ecstatic to see Canadians, and, possibly, women.

The *Petrov* just sailed from Hamburg in 20 some days. The crew is new to the tourist operation; their former work was on a Russian scientific ship. I didn't realize this is the *Petrov's* maiden tourist voyage.

I've enjoyed travelling with Wendy—she's great and adventuresome.

Wendy and I headed for dinner. I was deliriously tired and worried about our provisions.

The food and wine were disappointing. Haven't found the good Ushuaia restaurants yet. Two ship staff were at a table beside us and three others at another, dining with the captain and first mate. We were shocked when the man selling red roses pushed two onto us. I looked over at Tomas's table and he smiled. Tomas—the macho Polish-Argentinean penguin specialist. Who sent us the flowers? Adorable and ridiculous at the same time. Then Andy walked over to us and said, "Do you ladies want us to chaperone you home?" Was he serious? Is it still 1900?

We walked back with the slew of them. One Russian guy said to Wendy, "Good night. And good feelings."

I am so sick of purchasing items for the camp. I know it is important but I am aware of the paradox we are setting up. The Russians are starved for money, there is no longer much money for Antarctic sciences. Keeping the bases is also about political presence.

A few Russians tell Lena she has a good Russian accent. She laughs. She has too many responsibilities. She has $300 to purchase fresh food for the Russians tomorrow as their ship with provisions is delayed.

I want to build a home out of recycled things.

I respect the Russian ship crew member they call MacGyver. "He can do anything with a comb." Pyotr made aprons from the leftover linen for the ship tablecloths.

To buy, to do:
1. extension cord
2. transformers
3. double-check volunteer arrival dates
4. note for Bill Davis (ship expedition leader)
5. blue bag for *Petrov*
6. radios/charger
7. tape deck
8. more pillowcases
9. check provisions

DECEMBER 8, 1995

WT   Flowers and bushes in bloom. One minute we are in sweaters and jackets, the next
T-shirts. Food arrives to ship three pages short of my original dry goods order. Thank goodness Kevin, the chef on the *Petrov*, is willing to help out. Arrange to meet with our supplier
Gualdasi tomorrow before we set sail to see if he can send the missing items with our perishables on the *Multanovskiy*. Head to a supermarket to confirm translations and then a late
dinner with Dave and his new girlfriend. Black sea bass en papillote, an Argentinean white
and good company. We catch up on tree-planting gossip and struggle through the purchase
order. What a relief they both speak Spanish. Return to ship. Sleep.

DECEMBER 9, 1995

WT   Provisioning fiasco continues. Kevin, Lena and I meet with Gualdasi in his family
home. Serves coffee, freshly squeezed OJ, and we sample a succession of products paraded
through. Ordering is a language lesson and diplomatic challenge. What is dulce de leche?
How do you serve it? Sure, I'll try it. Leave feeling confident my food order will be filled
somehow, someway, but aware I'm going to be faced with some creative challenges over the
next few months. Back to ship. Launch after phone call to Mom and Dad, unable to reach P.

CD   We set sail at approximately 5:30 p.m. The weather is wet and a bit chilly.

It is a slightly sad yet romantic moment when the pilot comes aboard the ship to take us
through the Beagle Channel. Soon we'll be on our own in the Drake Passage.

We see Ushuaia fading in the background. The most southerly town in the world is out of
sight. I have a lingering feeling of fear mixed with anticipation we are travelling to a remote
place via the roughest ocean in the world and will be left on the other side.

For lack of energy to explain the concept of the Russian-Canadian environmental expedition to the Russian crew, Lena now describes Wendy and me as scientists.

I am mellowing out. Because of a schedule change I have no control over, we have four
extra unexpected days until Lena, Wendy and I are dropped at Bellingshausen. We get to see
more Antarctica with the tour group.

DECEMBER 11, 1995

CD   The Drake is relatively smooth: the Drake Lake, as opposed to the Drake Shake last
season. In a hurricane-like storm in the Shake at night we rolled in our bunks. I liked it, but
many were seasick.

I don't want to return to Toronto. I like working outside of the office. Being here is already
refreshing for me. I wish I could stay longer. I am reminded the world is full of possibilities
and how lucky I am.

I don't miss K much. I mean, I think of him lovingly—he's got good zest.

Many days on the ice soon.

I like finding out new things about me. I must take better care of myself. I know I should
live on my own, stop seeing K probably, and get myself into a really good environment.

Whales—I wish I could see a whale. I have a headache and greasy hair.

Notes from session with Gerry Spiess, motivational speaker, on the tourist part of the trip with us, not the project: After two failed solo sailing trips, he broke a record sailing the world's smallest vessel across the North Atlantic: a homebuilt ten-foot boat he likened to a 'sawed-off pumpkin seed'. Spiess studied Scott and Byrd to understand failure. He planned for a sixty-day sail, provisioned for ninety and arrived in fifty-four. He said the biggest danger was hallucination, being alone. "We talk to ourselves but say bad things. But we don't say bad things to other people so I created other 'friends' and held discussions. Otherwise I could have died from despair. The brain is fascinating. I created a third person so the two of 'us' could talk about 'him'; it was very helpful."

Gerry on Leadership:
· Shared vision
· Coaching—giving part answers, delegating
· Energizing the people
· Conflict resolution and team building
· Stop pampering
· Pay attention to details
· Face confrontations and don't fear different opinions
· Speak your mind
· If you think it, do it

WT   The ship is a big cradle rocking us to sleep. Some passengers appear only for meals and then pad back to their cabins to nap. Outside it's just water, sky and seabirds—we're all waiting for the first iceberg sighting.

Enter calmer waters today at Deception Island, old whaling station. Desolate, beautiful and historic. I wonder about what we will find at Bellingshausen and its relative historic importance; what we have the right to remove and what should remain. To bed thinking of a million things, waking regularly, wondering what the continent will be like tomorrow—my birthday. First exam dream.

DECEMBER 12, 1995

CD   Felt blast of Antarctic air when I went on the ship deck at 6:30 a.m. Only two of us were out there. On the *Petrov* they put the coffee on early. I couldn't sleep. Antarctica! Silent and serene. An oblong block of the whitest ice floated not far away. I thought it might be thousands of years old and maybe I'm the only person ever who saw it in that form. I studied the other bergs and when I looked again at the ice cream berg it was aquarmarine, reflecting sky, sea and the sunrays. I couldn't stop smiling. That iceberg was secretive; perhaps I could only see a tenth of it. Antarctica goes on impossibly like the forever horizon. I don't believe in heaven but if it exists, it must be this, upside down.

This ship is our miniscule spaceship in the massive ocean, our trip through the Drake Passage to the Southern Ocean like entering the moon's orbit, only horizontally.

Humpback whale fluke, Southern Ocean, Antarctic Peninsula, 1995

Abandoned whaling station, Deception Island, 1995

A wonderful night too. Except for Tomas taking it as a personal affront that I got out of his Zodiac tonight before he gave word. He can't understand the high I had from setting foot on Antarctic land. Surely he's not grown too used to the wonder of Antarctica.

I am pleased with almost everything. But I blew up the slide projector. The bloody expensive transformer didn't convert the current as it should have. I carried the steaming transformer to the lower deck and a Russian guy said he'd show it to the chief engineer. If we don't get it fixed the lectures for the volunteers will be dry—not good.

Wendy, Lena and I had fun in the cabin sewing expedition patches on our team coats. The engineer in crooked glasses came and said he was sorry he couldn't fix the projector. I was sorry he was sorry. Lena said at least we could use it as a table for our thread.

WT   Brilliant birthday. 33 candles.

8:00 a.m. wake-up. Present from Lena before I'm even out of bed and then Tomas arrives at our door with another. Just what I need: a poster of Brad Pitt.

Sun and first landing on the continent—Neko Harbour. Built a snowman, made snow angels, sat on a rock. So surreal—more monument than place. In a less hostile climate these mountains would be ski runs and I'd be in the lift line with everyone else.

Another touchdown on the continent at Paradise Bay. Lay in the snow, gazed at the sky. Caught a glimpse of an iceberg being born—I always seem to just miss them. Zodiac ride through cocktail ice back to ship.

Argentinean guests to ship for dinner. Yummy lamb with chickpeas and sesame sauce. Cake and champagne. Sauna and shower. This ship has all the mod cons—dried my hair today with a blow dryer so no fear of frozen braids. Drinks at the bar and now finally to bed.

1:00 a.m.: It's still daylight and I like to watch the icebergs out by my porthole. Tonight we drift; motors will start in the early a.m.

DECEMBER 13, 1995

CD   No idea of what day it is. A sailors' party for Wendy's birthday last night. Ouch—my head. This is a cool ship and the crew is proud of it and make it a special, safe and good place to be.

My first whales. Enormous humpbacks. So close. I shuddered at the sight of their serpent-like roll out of the water. Frightening at first. I felt like a mouse. But then I recognized their flukes. Beautiful animals, unbelievable and magical. Tender species. Free.

Lena said on a ship she feels like a grain of sand with the great sea all around her.

I fell in love with Antarctica last night. I knew I was in love when around 11:00 p.m. I stood on the deck and wanted to be alone in Paradise Bay.

There was the small, warm touch of life: the resurrected (after a fire) Almirante Brown station with a single light on. Above the minute spots of buildings were mountains crusted with a mauve glacier. Someone tried to talk to me but I needed to be alone, like I do now. I felt the vastness and the power of this bay. The mountains, the icebergs a testimony, a reminder of what we can't possess, only respect.

I felt so happy and still do. Again, like my subconscious mind brought me to the right place in time. I don't want to leave that feeling of connectedness to the mountains and bay but the moment has passed.

I think K's arms aren't big enough for me.

WT　Spring weather is more like winter today. Squalls and pack ice thwart passage and landing plans, but Carol, Lena and I are happy to savour the comforts of home while we can. Picked up freshly laundered and folded clothes below deck and then spent time in the ship's hold checking our inventory and bartering foodstuffs with Kevin. I know I'm obsessing, but I want to hit the ground running when we land.

Carol's lecture generates interest in the project. Many are envious, especially the crew who love Antarctica. Presentation on explorers was fascinating: what really has changed since the 1820s? Many of the dangers are still the same. Perhaps that's why I keep having the exam dream. I'm anxious to get moving. All this fuel and nothing to wear it off on.

Clouds lifted at 6:00. Sun. Landed on the continent for the last time at Charlotte Bay. Carol and I hiked up a rise in thigh-deep snow to lie in the sun and look out over the water. Slid and rolled and somersaulted down. Two penguins came up to us—they don't know the rules. Caught last Zodiac ride back to ship. I can't wait to get to Bellingshausen.

DECEMBER 14, 1995

CD　Tonight, glory around me in the form of water, ice sculptures and sun.

Ten days until Christmas Eve. I think it is Thursday although the day of the week has little bearing other than meaning we must leave the ship for the real work of this project to begin: pre-camp at Bellingshausen.

DECEMBER 15, 1995 // BELLINGSHAUSEN

CD　I am living on land. Major culture shock yesterday, transported not only to Russia in a matter of seconds when the Zodiac dropped us off, but to what felt like Soviet Russia. A most adorable thing: as we were about to leave the ship Pyotr helped us with our boxes and then asked where were our potatoes, "Gdye kartofil?". It was inconceivable to him an expedition would not have potatoes amongst its provisions. He rushed to get us some from the ship's cook.

Lengthy goodbye from MEI staff and we feel sad. The guys do three circles in the Zodiacs, say goodbye again on the radio. It seems as if news of the valley was there are women at Bellingshausen. Reportedly there had been no women living there for 26 years.

I did a quick walk around the station and drew a diagram. Debris everywhere, large and small, next to buildings, in the mud, piled on the beach and in a terrible landfill. Chicken bones. There will be plenty to do. The amount of debris is overwhelming and makes me realize that the Russians have a long-term project and expensive task ahead.

There is an international glacier study going on near us on the island. Scientists from Brazil, France, Uruguay, Russia, Chile and Argentina. The glaciologist from Brazil tells us, "The season is harsh, and the summer is taking a long time so the project is delayed."

Met a Brazilian journalist who is interested in what we are doing here.

Met two Sashas, Ilya, Sergey, Maxim, Yuri and others.

The guys were in the mess hall. It smelled like onions. They were preparing to vote in the Russian elections. They sat and peeled potatoes together while the Russian doctor, Sasha, explained voting procedures. Polls closed at noon and they sent a telex to Saint Petersburg with the results.

Sergey asked where I learned some Russian; I said near Sochi. He asked if I was a ballerina there. No; I picked plums near the Black Sea. "Nyet, ya kolkholsnizta," I said. ("No, I'm a collective farm worker.") He laughed. Gets them every time.

Cool but sunny good weather. Sergey got mad when I said I didn't need a hat; I'm from Canada. Pretty Czech boys disappeared on Nelson Island glacier, he replied.

The Russians have a one-year supply of food here. I'm embarrassed our big store is to arrive tomorrow and will be replenished along the way. We won't have more food than them but we'll get fresh food more frequently and we're going for (Antarctic) gourmet. I am stressed about the weather and hope the ship comes okay. Positive thinking.

One case of our beer appears to be missing. There is nothing to say.

WT   Happy Birthday, Mom.

Finally we are at Bellingshausen. Cookies and care packages from the ship's kitchen. Excited at the sight of snow and the promise that I'll be able to use my skis. Our own iceberg in the bay and penguins too. Mosses, lichens and rocks. The sun shining, beautiful, but through the mist it is moon-like and ominous.

Greeted at the beach by a military tractor to carry our provisions to a barrack-like building. Aren't actually allowed to help—just like on the ship. Taken to our temporary bedroom in the hospital, obviously the nicest room available.

Buildings are well-constructed and warm. Main building with mess, kitchen, and a variety of other rooms and diversions: a film room with an expansive collection, library and games room with a ping-pong table and dented pool table. Décor is sixties, but seems like a lot of money put into it initially. Transportation—all military. Concerned when we first tour the complex—wondering where people will sleep and how we'll function sharing a kitchen and dining room.

Fears abated later when Sergey gave us the building up the hill for our headquarters, and my sleeping quarters too. He was sweet to name it Canada House. It's charmingly rustic and steeped in Soviet history. The room best suited for the kitchen is windowless, with no hot water, zero counter space and dusty-rose walls, of all colours, but it beats cooking in the tent Carol scored from the Canadian government. A secret room hidden behind the sliding shelf in my kitchen will make a great pantry once I clear out the old radio wires. Everything is coming together; now I can put my mind to enjoying this place and our project. I just have to sort out where to store food so that it's closer at hand. Walking a quarter of a mile to get eggs in the morning is not going to fly. Language doesn't even frighten me much anymore. It will come. French and Latin 101 are going to be my saving grace.

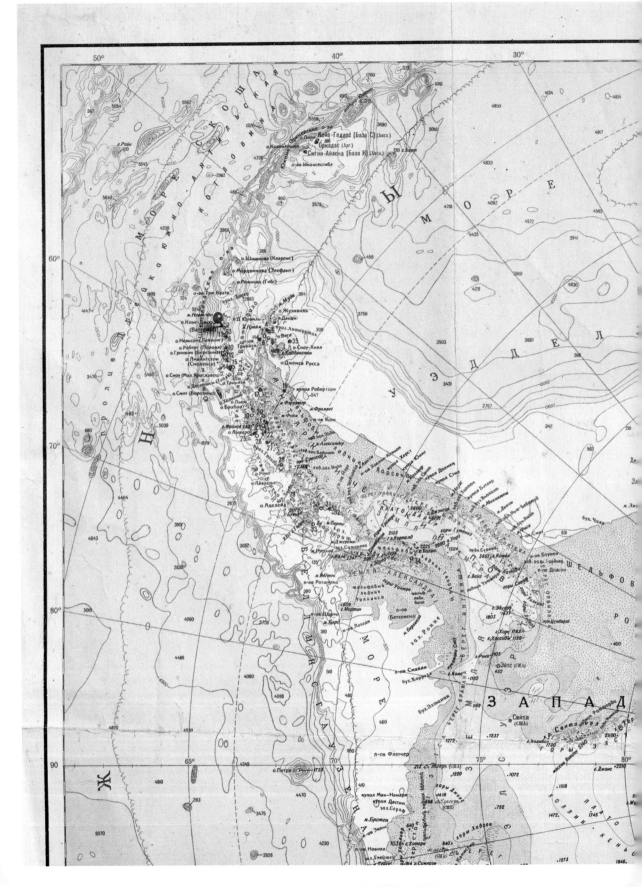

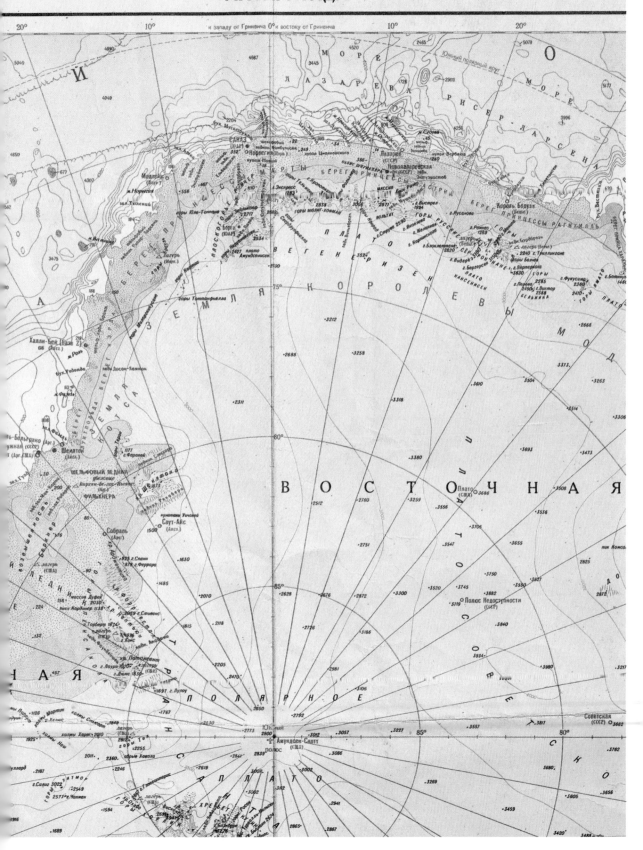

Wendy's cold storage shed behind Canada House, 1995

Volodya Cook scooping flour for bread, 1995

Looked through the Russian cook's kitchen after lunch. He's very cook-like in his whites and his kitchen a no-nonsense room: nice enamel cookware, boiler and hot plate running constantly, large industrial mixer and bread ovens. Menu what I'd expect of a camp diet: noodles, meat, broth, root vegetables and lots of spuds, tea and coffee—but the diluted fruit compote to drink is a surprise. Dry biscuits and jam—jam in everything.

Tour of surrounding area this afternoon—discouraged by garbage buildup and open burning. Meet Brazilians who are part of an international group of glaciologists camping on the glacier two hours away. We've been invited to go there in a couple of days.

We're spoiled. Every place we stop we're offered hot drinks and sweets. Vladimir and Ilya served coffee in their Russian Bear Den. It's by far the coolest building we've seen—high ceilings with a loft and a massive whale rib tied to the balcony railing. Both meteorologists. Vlad studies satellite photos of sea ice and makes a new ice map every ten days. Says he led the first American ski tour to the U.S.S.R. in 1975. Wants to be tutored in English and show me places to ski.

And tonight Sergey suggested I stay on here to name the mountains of Antarctica.

I have this feeling, a strange sense of something unfolding, opening in front of me. Bergs in our bay and bergy bits on the beach just in today with the strong wind—it's going to be neat to watch how things change. It's so quick.

10:55 p.m. Still light out—why did we bring flashlights?

11:00 p.m. Carol to me—feels like we've been gone for two years now.

DECEMBER 16, 1995

CD   The phone woke me up. I knew it meant the *Multanovskiy* was here and, I hoped, our food and equipment. Blue sky, warm day, blue ship on the bay. Wendy and I ran to the Zodiac driver and asked him to take us to the ship. Bill the expedition leader is friendly. We discussed cargo. I was pleased he said he'd love to do what we are doing.

We waited on the ship for the last items to be sent to shore, then, like skuas, scavenged from the ship's bar. One case Moosehead, one case Quilmes beer, one case white Piat D'or and one bottle of Grand Marnier. Wendy scored food in the meantime from Marco's galley— eggs, pork and phyllo to have something to make for the first volunteer groups.

There on the shore to help us was one Russian guy in the Cartier shirt. We don't know his name yet. At the storage room were eight of the Russians and Lena, waiting for my instructions about what to do with the ten massive unidentified orange crates loaned by the Canadian government Arctic program. They contain tents and a stove. We'll unpack the stove. The Russians are helpful and kind.

Have to be careful not to be an environmental narc. Not a hypocrite. Some of the Russians are feeding the skuas. I am tempted to criticize but I hold back. The soft sell works better. I pick up a rolling beer can. Lena says, "That will be the best example." We don't want to be a band-aid and do futile things.

To work on:
· recycling process/waste disposal
· covering food
· not feeding skuas
· returning trash to Russia or somewhere with good disposal practices
· stop open burning

It is a privilege to live here, get insight into the scientists' and staff's Antarctic life and routines. Meals: tasty unfresh potatoes, zucchini, meat.

We go to the radio house where Sasha makes us coffee. Lena pours it out. At lunch we are reprimanded for not wearing enough warm clothes although we are fully suited up. The Russians are going to care about us no matter what because we are women, we're young and we're their guests. They are incredibly generous to us; when they learn about our provisioning problems they offer us surplus staples. Noodles, sugar, flour.

Soon we must visit the Chilean and Chinese bases.

I met with base chief Sergey Potapov. Lena helped translate. The table had a flowery doily and on it two ashtrays. I took a cigarette he offered—Argentinean—but couldn't smoke it; too strong and it was 9:00 a.m. Sergey is kind and concerned about our well-being and that of our groups. I explained our goals and thanked him for the chance to work here.

We couldn't manage without Lena. A few younger guys like Dima know English so we converse a bit but most don't speak more than a few words of English, like Sergey.

He showed us the sauna and shower room, recently built and in fine condition. It's available to us every day but Sunday.

Sergey said we can use their vehicles for work and excursions but the amphibious tank they call the PTS ("plavayushchij transportyer sryedniy") is in demand and we'd have to let the drivers know in advance. We said we didn't want to use the vehicles except when necessary.

The scientists at Bellingshausen are doing:
· *Aerology (upper atmospheric studies)*: daily balloon launch at approximately 6:15 p.m. Temperature, humidity, wind direction and speed monitored throughout flight. Altitude gained varies between twenty and thirty km.
· *Meteorology*: six hourly observations of temperature, wind speed, cloud base, precipitation. These results are compiled and transmitted to Moscow.
· *Sea ice monitoring*: Four or five times per day infrared satellite imagery is recorded showing coverage and density of sea ice in the area of the peninsula and continent. Every ten days results are compiled and mapped.
· *Seal population and distribution monitoring*: A weekly count is conducted on the Fildes Peninsula area throughout the summer season. Elephant seals are the main source of interest. However, all other seal populations are also recorded.

After our lunch it was women's sauna time. It was lovely, a sauna and shower. Wendy ran to the stream and jumped in! We walked to Frei to use their radio to call Ushuaia and make sure the volunteer flights and ship schedules are okay.

Lena and I planned a site orientation for the volunteers and did a waste audit. We found:
· welding rods
· scrap metal
· nails
· paint spills
· cloth
· plastic
· cigarette butts
· abandoned fuel pipes, fridge and metal objects

We decided we'll start with the pipes at the abandoned fuel tanks as that could take time and we'll get it out of the way with the first groups of volunteers.

We got their rooms ready in the block next to the Russian's main accommodation block. Sergey gave us five empty rooms to use. The rooms are well-heated, austere, clean and furnished with two single beds. Shared bathroom. It's clear in the station's heyday there was capital for the base. In each room we put a brochure "Guidance for Visitors to the Antarctic" and an info sheet on safety, water conservation and recycling.

Waste plan in accordance with Annex III of the Scientific Committee of Antarctic Research (SCAR) Bulletin, July 1993
· All our waste is returned by Marine Expeditions vessels to Ushuaia for further disposal
· Waste materials are centralized and divided and bagged. Burnables: paper, cardboard; plastics: bottles, pvc packaging, sweets wrappers; organics; glass; metals: drink and food cans, silver foil
· Human waste is disposed of using Bellingshausen's existing facilities

WT   Big day. Concern over our lack of provisions. Must be confusion about which ship will deliver to us. Out to the ship with Carol to see what I can pull from their hold. Marcos is obliging and provides me with whatever I need to tide us over.

Sasha radioed the *Petrov*, but it was out of range. Eventually got through to the radio operators in Moscow to ask them to contact the ship. Message later from Tomas on the *Multanovskiy*, "Yes, perishables will arrive on the *Petrov*." Until then, stone soup anyone?

Funny scene unloading the PTS. The ship's crew and Bellingshausen team put as much energy into preventing us from lifting as they do into moving crates. I feel a little useless, but they seem happy helping. We follow Lena's lead and don't mess with the rhythm trying to prove ourselves.

Wendy in her kitchen in Canada House, Bellingshausen, 1995

Sergey assigned the Sasha that lives on the hill to be my go-to guy for set-up and this evening Lena reported he waited all day by his phone for me to call with a job. And Doc Sasha ordered us not to work because Lena has a cold. I'm not sure I'll get used to being on someone's radar like this, but the warmth and playfulness in their eyes makes me feel glad to be a part of it and fortunate to do what I do.

Relieved when Sergey gave us a tour of their provisions and offered me anything I need to make our camp work. It's dark tonight and I think I'll be able to sleep.

Ideas:
· Shoot portraits
· Whalebone/landscape/berg drawings on weathered boards with nails from Bellingshausen—art lesson—environmental lesson—composition/prints/nothing
· Map of King George Island
· Ice map
· Satellite
· Replica of secret door
· Photographs of interiors/equipment, vignettes—smoking/games/potato-peeling
· Crates
· Seal pathways
· Penguin prints
· International opening—foods
· Canada House—white paint embedded with thread
· Inukshuk

DECEMBER 17, 1995
CD   A brilliant morning. Sun turns berg in bay into gold. We had breakfast early with the Russians. Hot bowl of kasha—earthy-tasting porridge.

I am sitting on a rock. A skua is begging, waiting for food. This morning Sergey fed the skuas chicken legs. Bad. What can I say? He knows it is wrong. But the bird is his friend.

Lena asked me for tomatoes.

We went to the base camp of the glaciology project. Invited us to their kitchen tent for coffee. A handwritten label on the sugar lid in four languages: *caxap, zucker, sucre, azucar.*

The glaciology study is on hold. Dead equipment batteries. When the weather is good they work day and night. Well, there is no night. There are some dangerous crevasses near the extreme camp, closer to Arctowski.

It is cold but sunny. Never get hot, never get cold. The Inuit know how to stay the right temperature. Gerry Spiess said this.

Everyone here pulls his weight and is busy. I'm surprised at the amount of science going on. What happens to the results I don't know, but these guys are working hard. While some research seems tedious and archaic with the men sending up meteorological balloons at 6:30 every day of the year, these guys take their roles seriously. Everyone takes shifts: Sasha the satellite operator sleeps five hours per day. The two scientists, Vadim and another

Sasha, do the daily meteorological studies. Everyone does something of everything: dish duty, working a day in the kitchen.

We went where biologists Dima and Vlad were counting and studying elephant seal migration from the Drake Passage to the Bransfield Strait. Although the seals could swim around the Fildes Peninsula, they migrate through the station, slithering, heaving up and down the station through the stream, past rejected pineapple rings, into the sea.

We drove up to Volchevno, which means something like The Magical Place. The tank carried us up and down hills. Sasha was deft manoeuvering it. I thought about the Greenpeace report of tracks everywhere. But this is economics; this is the vehicle the guys have for now, besides Sergey's little car, the Niva. It fits many people and is all-terrain.

"Look left," said Sasha the mechanic.

I popped open the tank's door. The side was painted with the word Russia in Cyrillic.

Overcast. Windy. Spied humongous elephant seals. Then smaller, smoother ones, still big. They had doglike faces, scary mouths with purple tongues.

The seal colony. They stared at us at first but carried on as if we were irrelevant. Scratching their "arms" with their fur-covered "hands." Two seals were hugging each other. One put its arm over the other's back and made like a kiss. Then some seals scrapped—males with teeth-marks in their skin, chopped-up fur. We are all seals perhaps.

We moved from the seal colony to a hut for the biologists. Another exquisite experience. The shack was a wagon-like trailer now held not on rocks, but whalebones! It was a shabby hut with green oil paint chipping off in big chunks—sundried cracks all over. Inside were two beds.

Nature mirrors nature. A rock sitting high on another rock looked like an elephant seal.

This is a lesson on minimalism. Every hut is a treasure, is useful. Recycled. How often does McMurdo, the big U.S. station, buy more stuff to take down? At least the machines here get full use and are kept in good condition because they are all the Russians have.

I realize I want to write about this experience after I leave the VIEW Foundation when I'm unattached to work politics. I want to move onto a new challenge but I am proud of what we've accomplished in a short time.

We had come to the hut so the Russians could fix the hinges on the door. I was cold so I went inside and sat on a bed, wrapping found blankets around me. The men hung the door loosely and smiled, closing me in for some time. I started singing "Milen'kij ty moy," a folk song I'd learned in Sochi. It had a male and female part. Two of the men joined in: me inside the hut, them outside.

"This is nice," said the biologist Vladimir.

On the way home I sat in the back of the tank. Meteo Sasha said, "Look out the window. China." There stood five Chinese men in red expedition coats against the white glacial backdrop. We stopped; Wendy and the mechanic/driver Sasha got out. I felt I was in China for a moment. The men were smiling and chatting, although I couldn't hear them.

When the Chinese scientist handed Wendy a chunk of petrified wood, she said she felt insignificant. It was from the time before Antarctica was ice, before the ice age. It was forty million years old.

We will leave the piece of wood here. This is where it belongs.

And day turns into day. The light now at 7:00 p.m. is like 2:00 p.m. light at home.

I had an urge to open my Christmas presents early.

Wendy and I have great discussions. I suggested ecology needs a human face. Greenpeace serves as a good watchdog for the planet, keeping expectations high and reporting on infractions of rules or principles. But Russia is in a transitional stage, Russia is poor, Russia was closed off for forty years.

Bellingshausen factors: 50 percent funding cut from Russian Federation government since perestroika. Station from 1968 hardly upgraded, in time warp running with old machinery. Well-built base then, strong buildings with high ceilings and thick doors like a ship. In them I still feel like we are moving. Neighbours China, Chile, Uruguay, Argentina, Korea, Poland, Brazil. Language is a barrier to developing a rapport with other stations. Economic inferiority complex; Iron Curtain didn't come down that long ago. Isolationism.

Late at night: Sergey told Lena that the guys told Maxim and Yuri that they had to "stay away from our girls." They had noticed them flirting with us. Group dynamics.

Lena told us about being at Novo station and cutting a man's hair. He was so stimulated he couldn't sleep for nights. She had to stop giving haircuts.

"Rules of Thumb" I found in Greenpeace papers:
· Stay at least 200 metres away from giant petrel nests
· Do not feed skuas or other wildlife
· Do not approach leopard seals; if you are not sure of the species, stay away!
· Try not to come between a fur seal and the sea, and do not approach fur seals (they can move surprisingly fast and their bites are painful and often become infected)
· Avoid stepping on all vegetation (including mosses and lichens) as well as on sensitive soils and soil features
· Make sure that all rubbish and other wastes (except grey water and sewage) are taken out of the Antarctic
· Stay out of Protected Areas (if in doubt, ask)
· Always give the benefit of the doubt to the environment

Dima, Hilltop Sasha, Radio Sasha, Sad Vlad (left to right),
Lena (in front) on the radio hut steps, Bellingshausen, 1996

WT   Clear, no wind, sunny, plus 4°C. Highest temperature yet this season, we are told.
Tractor ride to Stoney Cove. Moss a completely different colour today.
Found whalebone
Clean stove

Bellingshausen Team:
· *Sergey Potapov*: Base commander. Fifty-something. Looks like he has lived a little. Stern at times, as if he is trying to make up for being a one-of-the-guys kind of leader. Laughs easily. Kind.

Five Sashas (Diminutive name for Alexander):
· *Driver Sasha* (Sasha Diesel): Diesel Power station chief. Forties or fifties. Paul Newman eyes. Don't think he has much hair under his longshoreman's hat. Speaks a bit of Spanish.

· *Hilltop Sasha* (Aerology Sasha): Meteorologist. Lives with Vadim. Think he is 29. Has a three-year-old. Intense and on the case. He's making trivets for our dining room table. Speaks English. Actually most of the scientists speak some, at least the younger ones.

· *Doctor Sasha* (Doc): Fifties—could be older? Salt-and-pepper beard with crazy eyebrows. Seems a bit nervous, trying hard to make us feel at home. Tomorrow he is going to open the operating room to perform surgery on my stuffed Panda. Speaks English and Spanish.

· *Radio Sasha* (Sasha Radio): Unflappable camp communications guru. Forty-something. Leprechaun-like Ukrainian, if that's possible. Marital status unclear. Speaks some English.

· *Meteo Sasha*: Burly, keeps to himself. Hasn't said a peep to any of us, but seems nice enough.

Four Vladimirs (Diminutive name: Volodya):
· *Satellite Vladimir* (Sat Vlad, Sad Vlad): Meteorologist. Forty-something. Bearded, with bright blue eyes and an athletic build. The word is out about my stuffed bears—he confessed he brought a stuffed pig his 15-year-old daughter made for him for 1995, the year of the pig. Speaks a bit of English.

· *Biologist Vladimir* (Bio Vlad): Works with Dima; otherwise seems to be a bit of a loner. Fortyish, fit and very friendly. Has offered to teach me Russian. Speaks a bit of English.

· *Diesel Vladimir* (Volodya Driver, Diesel Volodya): Works in Diesel. Has been driving us around. Forties or fifties. Brawny and balding with a big moustache. Seems a little grumpy.

· *Cook Vladimir* (Cook, Volodya Cook): Fifties? Welcoming, salt-of-the-earth kind of guy with a shock of white hair. Gregarious and tall.

· *Vadim*: Aerologist. One of the camp elders. Grandfatherly beard and moustache hide his smile, but I can tell one is in there.

· *Vassiliy*: One of the support guys. Forties or fifties. Wears a classic Russian fur hat. Can't miss his gold tooth when he smiles. Does so often. He'll be in charge of delivering water to Canada House. Speaks Spanish.

· *Anatoliy* (Diminutive name Tolya): Support team mechanic. Forties or fifties. Dark, wiry and compact. Everything about him is quick. Speaks a little Spanish.

· *Dimitriov* (Diminutive name Dima): Biologist. The youngest member of the team—twenty-three, I think, Tall and lanky. Fresh-faced and very sweet. Speaks some English.

· *Valorah*: Ecologist. Quiet and gentle looking. Seems kind of vulnerable, as if he isn't sure what his role is now that we're here, but I could have it all wrong.

· *Ilya*: Meteorologist. My age. Dark and good-looking with a slow smile. Lives with Satellite Vlad. Think Carol has a crush on him. Speaks a little English.

· *Yuri*: Glaciologist with good tan. Flying out soon.

· *Maxim*: Glaciologist. Fifth time at Bellingshausen; has overwintered fourteen times. Second field camp. Flying out soon. Speaks pretty good English. Has a nineteen-year-old son and a fifteen-year-old stuffed mouse Olga, named after his wife.

DECEMBER 19, 1995

CD   Dead seal pup; seals moulting, sleeping and scratching; moss; lichen; rocks changing shape and size; icebergs, one looking like a seal, one a whale fin. Bio Vlad told a story of fur seals he worries about. The ice cliffs were up to nine metres high because of the late spring and he was concerned many fur seals would die because they couldn't get up to migrate, but they did.

At 2:00 we visited the Chilean base, Frei. Lena, Wendy, Doc. Zuniga, the base commander, is a military type. Big, tough-looking, but he didn't intimidate me. I felt we gave a good introduction, explaining who we were, that it was a joint project with the Russians and that we were going to start a cleanup project. I said we are not extremists, we want to learn and to help. This is good he said, since there are problems everywhere.

We had a killer drink, pisco sour, and I got a big buzz since I've eaten little in five days. Wendy scored the recipe.

Zuniga invited us for lunch tomorrow. Also to an air force party in the gym on December 22 for the flyover of Santa Claus.

Sergey told me the boys are changing their shirts for us. He'd seen them in the same thing for nine months and "now it is a fashion show." Sergey was tired of the guys asking him what they can do for us. He said, "They've gone a bit nuts over you," but he and Lena agreed this is okay because having us at the station is positive, it's a motivator, it makes them think of their women—their wives, kids, mothers. He's a sensitive, if maladroit, chief. He's sick now. I've given everyone my cold.

The guys know each step we take. They are very disciplined, odd, hierarchical.

I finally found time to read the Ursula Le Guin story from *The New Yorker* (1982) I brought, "Sur: A Summary Report of the Yelcho Expedition to the Antarctic." It's awesome. Sur is the fictional story of eight South American women (all crew, no officers) who travel to the South Pole, arriving one year before Norwegian Amundsen did in 1911, making them the first humans to reach it, but they refuse to publish their journals.

The narrator is angry with the mess in Captain Robert Falcon Scott's base camp, the spilled biscuits and tins lying around and says, "Housekeeping, the art of the infinite, is no game for amateurs." She loses her toes at the Pole and is happy she leaves no footprints.

WT   More typical King George weather. Cloud. Wind. Way too tired to write much.

Home from hike to the Drake and Flat Top in time to load the lorry for Canada House. Man, I love their Queen's English, especially Aerology Sasha's—"lorry" describes that dependable old beater perfectly. Six helpers made for short work and all of them tried to teach me Russian: Ilya, Hilltop Sasha, Doc Sasha, Vassiliy, Biologist Vladimir and Sasha Radio. Break for lunch, yummy soup. Discover a feather in my hat. Vassiliy confesses. We communicate partly in English, partly in Spanish and the few Russian words I've learned. Can't believe these guys and their generosity. At dinner Bio Vlad gave me flash cards he made and a Lenin pin. Pins from Doc Sasha as well. This must be what it's like to be one of the popular girls in class.

My kitchen is shaping up. Sergey gave me a massive pot to heat water, enamel basins to make a triple sink, buckets for recycling, and I stacked some of the wonky veneer cabinets for storage. Fridge and stove arrive tomorrow. Was working tonight when Maxim dropped by with chocolate, Jim Beam and a *Moscow Times* mug. The chocolate was delicious, but I passed on the mug of whiskey. Happy I did, as I'm conscious of the scandal it could create.

He was good company as I set up my pantry. We got into a conversation about how to distill vodka when I told him my plan to collect recipes from all the bases. I asked him if he knew of any edible wilds on the island and got a recipe for the cookbook: sea cabbage salad made with laminaria (fresh kelp). Maxim also thought of a recipe from the glaciology camp: caipirinhas made with whiskey and lemon instead of cachaça and lime. Love the way he wrote it—he is specific about the brands: Jim Beam or Ballantine's. No doubt he knows his recipe is more whiskey sour than caipirinha, but I guess mixology rules are flexible when stores run low. Soothes the Brazilians as well.

Got to sleep now.

DECEMBER 20, 1995

CD   We headed down the hill at eleven o'clock to get ready for our special luncheon with the Chileans. When I stepped outside Canada House I saw two Brazilians from the Glacier camp. Also Roberto, the Uruguayan doctor and Doc Sasha: my international neighbours, standing on the black rocky beach on Maxwell Bay, chatting about Antarctica in the sun. I joined them. I love this job.

Lunch at Villa Las Estrellas, Habitacion 51. We went into a special room and were offered a pisco sour (again). It was a pleasant room. I sat beside Commandante Zuniga. We had good

Vadim launching the daily balloon from the aerology lab, Bellingshausen, 1996

chats in Spanish. Then fourteen of us moved to a dining table with a beige tablecloth. I chose to sit beside the Commandante. We discussed Tibet. I told him about the book I wrote about Tibetan women in the independence struggle. Zuniga said you can't write about something of which you are not a part. I disagreed, and agreed.

Someone apparently praised Pinochet. Did I hear that right? The guy at the table across from me caught my eye—as if we agreed Pinochet was a murderer.

We started lunch: fresh artichoke dipped in mayonnaise with lemon. Delicious.

Then a meat soup from which I ate a potato and piece of squash. Thank God I'm the only vegetarian so it's not a bigger hassle for the cooks. White Chilean wine, herbal tea from the Andes (fashionable in Chile I'm told), and a sliced orange.

Doc toasted, "To new friends and new ideas."

Zuniga said, "Thank you," and raised his glass.

On the walk home we tried to talk Doc into being Santa Claus. He won't do it; he doesn't feel right about it. He's a beautiful person. I want to give him a present.

The Russian guys partied today while we were out.

Lena and I had an excellent conversation tonight. She knows so much about Antarctica. She's a very cool woman. I was worried she'd be in an awkward position as go-between but she's a pro. We get on well, I think. I wanted to find out how she perceived our presence. Lena said all she hopes is we clean the beach and take out some garbage. What we can do is minimal. She said a proper beach cleanup could take us the whole season. I am ambitious but I appreciate that our goals must be realistic. She said she cleaned a 4 × 4 m piece of beach for an entire month. As far as she is concerned, where we are is not Antarctica—it is too crowded, too full of tourists, so highly altered.

I tried to persuade her we shouldn't give up on it.

Money for cleaning up, she emphasizes, is key, and the Madrid Protocol.

The bay was still and beautifully purple-hued tonight as we walked down the hill at 11:00 p.m. There was a blue slit of sky, a brown rock shining in the calm violet-navy water and lumpy shaded icebergs.

WT   Nap time, but day so full of events I want to record them now. Volodya Cook made bread after breakfast. Think I got some decent black and white shots of him at work. He is definitely of the pitch-and-toss school of cooking—he had the dough thrown together and rising in no time. Had to watch his gestures carefully to follow the recipe.

Confirmed the ratios with him afterwards. Don't think I'll be making the version calling for 100 cups of flour soon. Later, when he was pulling the loaves from the oven, he showed me a trick to protect the bread from drafts as it cools—place a tea towel overtop. Volodya says (or at least Lena says he says) a damp cloth draped over the crust of day-old bread will keep it moist.

At lunch we had borscht. Cook served the last of the old bread and reluctantly gave us each a slice from the fresh loaves. I'm glad Lena pressed him—she has no shame. Sure, it's hard to cut when just out of the oven, but I wasn't buying his silliness about bread being bad for you when warm.

MAXIM'S MOONSHINE

Lena's translation: "I am afraid neither you, ladies, nor I would ever make it since it is for home-brew alcohol, absolutely disgusting stuff, which may be consumed at an Antarctic station when nothing else is available. The recipe ingredients include yeast, sugar and water with no indication on the amounts and proportions. I guess one should make a decision himself depending on the amount of initial material at his disposal. Apart from yeast and water a distillation apparatus is needed. And the last detail I should add is that the Russian name of this drink is *'samogon.'*"

SEA CABBAGE (LAMINARIA) SALAD

I can't imagine following Maxim's recipe to the letter creates a real crowd-pleaser, but maybe if you cut the pieces of kelp small enough and round out the textures with lots of red cabbage.

Collect living kelp.

Clean with fresh water.

Cut.

Add onion, garlic, mayonnaise.

CAIPIRINHA

As described by Maxim, International Glaciological Expedition '95

3 ounces whiskey ∥ ¼ of a lemon ∥ 1 to 2 tablespoons of sugar

"Not mix, but
  mix with press,
  and ice of course, and drink is ready.
  Attention…Ready…Go!"

[*Translation*]
Start with a freshly washed lemon and roll it around on a smooth surface to loosen the juice before cutting out a wedge. Place the lemon wedge in a glass, pulp side up, and sprinkle with sugar. Muddle the lemon with a pestle just long enough to release the juices or it may become bitter. Add the whiskey (2 ounces is probably enough) and stir. When the sugar is dissolved add ice (crushed, cubed, whatever you have) and stir again.

Makes one drink.

PISCO SOUR

Pisco is a brandy made in Chile and Peru, and both countries lay claim to the pisco sour as their national drink. Distinctions can definitely be made: is it lime or lemon, simple syrup or sugar? I can't recall which of our Chilean hosts gave me this recipe, but when I checked the original in my archive I noticed the words "Esta es su casa" written across the top of my notepaper. I like the idea of serving a cocktail to guests that seems to say, "This is your home."

4 ounces pisco // 2 ounces fresh lemon juice //
2 egg whites (or 4 teaspoons egg white powder) // sugar to taste

Combine the pisco, lemon juice, egg white and sugar ($1\frac{1}{2}$ tablespoons is a good place to start) in a cocktail shaker full of ice. Cover and shake vigorously for about 15 seconds, taste, then make adjustments to the sweetness and shake again. Strain into chilled glasses and serve it straight up topped with spoonfuls of lemony foam.

Makes two drinks.

COOK'S BREAD

We didn't share a language when the Russian cook let me help him make bread, so he mimed his way through the method using whistles and exaggerated gestures for punctuation. As the bread was baking he showed me a chart so I could confirm ratios. Some day I'll try the version that calls for 100 cups of flour.

7 tablespoons active dry yeast // 2 cups lukewarm water //
2 tablespoons of sugar // 1⅓ cups butter // 6 tablespoons salt // ½ cup sugar //
8 cups hot water // 27 to 28 cups all-purpose flour

Whisk together the yeast and 2 tablespoons of sugar in a medium bowl containing the lukewarm water and set it aside until frothy.

While the yeast is proofing, put the salt, butter and ½ cup of sugar into a large bowl (a large salad bowl will work nicely) and pour in the 8 cups of hot water, stirring well.

When the butter has melted and the mixture has cooled so that it is warm to touch, stir in the yeast followed by about a third of the flour (around 9 cups). Gradually add more flour until you are no longer able to use a spoon to mix, then dust your hands with flour, coax the dough into a soft ball and turn it out onto a floured surface. Continue working the dough gradually adding just enough flour to keep it from sticking to the surface. You'll know you've done good work when you feel the dough becoming smooth and elastic. This could take 10 minutes or more—the longer the better.

Return the dough to the bowl after greasing it with butter, turning to coat the dough all over. Cover the bowl with a tea towel and put it in a warm place to rise for about 1½ hours.

To make the bread

When the dough has doubled in size turn it out onto the floured surface and bring it down with a few punches (let your mood determine how many). Leave it to rest for a minute before continuing to knead, fold and punch it a few more times. Divide the dough into the desired number of loaves. Form each portion into a ball and set aside, covered, for about 10 minutes.

To make the loaves, use a rolling pin to roll out each ball into a rectangle slightly longer than the loaf pans you are using. Starting with the narrower edge, roll up the dough like a jelly roll, carefully piercing any large air bubbles as you go. Finish by pinching the last seam together; seal the ends of the roll in the same fashion and fold these flaps under the loaf. Place seam side down into a greased loaf pan. Continue with the remaining dough, cover and set aside to rise for about an hour or until it has risen above the top of the pan.

Bake two or three loaves at a time in the centre of a 350°F oven for 45–55 minutes. You'll need to control the rise of the remaining unbaked loaves. If you live in a drafty house, place them close to where the cool air is getting in; otherwise make some room in your refrigerator. Bring one set out to rise at room temperature as you place another in the oven. If you are baking more than two loaves at a time rotate the pans twice; first after 15 minutes, then after 30 minutes. You'll know they are ready when they are nicely browned on top and sound hollow when you tip them out and tap the sides and bottom. Pop them back in the oven if you think they could use more time. When done, remove from pans immediately and transfer to racks to cool. This bread is very sensitive to air temperature—to prevent the crust from becoming wrinkled and cracked, cover your loaves with a tea towel as they cool.

Makes six large or twelve medium-sized loaves.

If you don't feel like baking the entire yield of Cook's Bread, wrap the remaining loaves tightly in plastic wrap just after you've shaped them and freeze—the dough will continue to rise a bit before it freezes and you want to prevent it from taking over your freezer. When you are ready to bake, unwrap the dough and place it in a prepared loaf pan to thaw and rise. I like to let mine rise overnight on the counter to bake for breakfast. Follow baking instructions as outlined above.

If you've signed on to the task of baking all your dough and are strapped for loaf pans, use what you have on hand. Cake pans, baking sheets, muffin pans; rectangular, round or square—all work well. I have even used tin cans and clay flowerpots. Estimate the baking time according to size and when in doubt, set your timer for an extra five minutes, take a peek after two and trust your eyes, ears and nose.

Scaled-back ingredients to make one large loaf of Cook's Bread:

$1\frac{1}{2}$ tablespoons active dry yeast // $\frac{2}{3}$ cups lukewarm water //
1 teaspoon sugar // $\frac{1}{4}$ cup butter // 1 tablespoon salt // 4 teaspoons sugar //
1 cup hot water // 4 to 5 cups all-purpose flour

"I was also convinced that if everyone was to remain in the best possible state of mind, our diet had to be close to the one which our countrymen are accustomed...The question of bread, so important to the French, was finally resolved thanks to the skill of our cook. Three times a week he would make fresh bread for us, and on Friday and on Sunday mornings he would even produce rolls made with butter. On other days we ate the galetas, made at Ushuaia, which were remarkably good."

—Jean-Baptiste Charcot, *Towards the South Pole aboard the* Français: *The First French Expedition to the Antarctic 1903–1905*

NATURE RULES

Carol

———

I love that science is an Antarctic currency and tool of diplomacy. Our civilian cleanup tried to put policy in action in this remote and once forested place.

A century ago Belgian, British, Norwegian, German, Swedish, French and Japanese expeditions rushed to Antarctica where few humans had yet reached. Roald Amundsen arrived first at the South Pole on December 14, 1911. Explorers counted on a blend of skill, daring, ego, adequate supplies of fuel and food and charitable weather. Most went for empire, but when times got tough food was king.

Antarctic exploration and pursuits of domination continued in the late 1920s, including by air. In 1939 over New Swabia, the Antarctic area the Germans explored and claimed, Nazi planes planned to drop dropped hundreds of steel spears bearing the swastika into the ice.

During the 1800s ships from several territories (but mainly the U.K. and the U.S.) exploited the region's seals. After those were wiped out, Norway and Britain began whaling and later so did Japan and the Soviet Union. Nations vied for Antarctica's resources and territory.

These attempts to claim and subdue Antarctica changed in 1957–1958 during the International Geophysical Year. Thousands of scientists from sixty-seven nations collaborated on a comprehensive scientific program, studying the earth's health, water and weather. Several of these nations made a stunning next step.

In 1959, twelve countries negotiated the Antarctic Treaty, declaring the continent "a natural reserve, devoted to peace and science." They did it largely in secret during the Cold War. The original signatories were Argentina, Australia, Belgium, Chile, France, Japan, New Zealand, Norway, South Africa, the Soviet Union, the U.K. and the U.S. The Treaty aims to protect the continent's fragile ecosystem and resolve territorial disputes. It was also the first nuclear arms agreement. Then the Madrid Protocol to The Antarctic Treaty in 1991 prohibited oil and other mineral exploitation for 50 years.

By 2012, 49 nations representing more than three-quarters of the globe's population had signed the Antarctic Treaty—a rare achievement in a world beset by conflict.

In 2041 Treaty countries could possibly renegotiate the ban on mining. The resource race in the thawing Arctic may forewarn us of Antarctica's fate. The future of all regions also depends how individuals, industry and politicians respond to climate change and consider the human and ecological impact of a new land and commodity frontier.

There are some 45 research stations operated by 20 countries with over 4,000 scientists and support staff based there during the summer and a growing number of ship-based tourists.

At times Wendy and I both thought maybe we shouldn't go to Antarctica.

HOME

DECEMBER 21–31, 1996

No one was daunted by hard work or the drab weather.
We kept pulling, cutting and bundling.
It got chillier, rain almost turned to snow. Job won't be done today.

DECEMBER 21, 1995

WT   First night in new home. Left blackout curtains open; fantastic light at 3:00 a.m.

CD   I got a call from Wendy over in the Russian kitchen; she said I should come to meet the base commander and pilot visiting from Artigas, Uruguay. Friendly, one balding, one with silver-black hair. We had typical chit-chat—where are you from, how many at your base—and I explained what we were doing. They had come to meet with Sergey and see his new price list for logistics services which we helped him make in English.

After breakfast Lena and I went to the Chilean base next door to post letters but because of the time difference in Chile, it was not open yet. We forgot the bizarre one-hour time change. Then we worked on Canada House: more interior decorating in the living and dining room. We wrote out the Russian alphabet to adorn the walls. Then I mopped in Meteo house and set up beds. I looked for the propane hookup for the fridge and walked back to Canada House. The volunteers are coming.

Lena said there is a Russian belief, "When you sleep in a new place for the first time, you will dream of your fiancé." She also said, "This is an expedition, you work as a team."

Vlad was hungover today. Everyone felt a lot of compassion. We missed him at our party. Want to give back to all the guys helping us all the time. Wendy prepared an awesome feast for all. When she served us the loaf of warm braided bread everyone looked at it longingly and then looked for a knife. I knew from Wendy's exhibition opening where she also served her bread that we were meant to pull it apart with our hands. "Someone rip off the first piece," I said, "that's what Wendy wants."

> Scientist Charles Swithinbank, on the first international (Norwegian-British-Swedish) Antarctic expedition, in 1949–52 said, "People have got more and more specialized now, whereas at the time we only had a few people and we didn't have all kinds of specialists…I was there as a glaciologist but a glaciologist needs a map to plot his observations and there was no map, so we were making maps at the same time… You don't have to be a genius to contribute to knowledge."
>
> —John May, *The Greenpeace Book of Antarctica: A new view of the seventh continent*, 1988

DECEMBER 22, 1995

CD   We've been preparing for today forever. The volunteers arrive. We can start the cleanup in earnest. I feel nervous but excited to get cleaning. But what is the best approach? And who are we anyway? There is no guarantee that after we clean up the stuff that the Russians will maintain or continue the cleanup. They are already recycling, meaning the way they use and reuse equipment.

Excellent to meet Sean and John. Now our staff is complete. Sean has a wry sense of humour, tall John is a bit awkward. John was saying how different yet the same Bellingshausen looks since he was here in 1968.

I was impressed with the first group of seven volunteers. They were well-travelled and bright. I was happy they were sincerely keen to work. Our Christmas group included a family from California—a fifteen-year-old boy with his mother and grandmother, Andrew, Jeanne and Marianne; a couple, Lou and Dewey; and two men, James and Bob. Bob is with *Condé Nast Traveler* magazine. He's part of the reason I've been nervous. He's writing a story on our cleanup.

The volunteers arrived off the ship a close group that laughed a lot. Lou said he felt it was privilege enough to be on land at a research station, let alone to visit other stations.

Bob is a cool guy. So eager to get Russian, Chilean and other stamps in his passport. Excited as this is the only continent with many countries and no borders.

WT   Late arrival of ship. Bedlam on the beach.

Three Zodiacs full of food coming to shore—crates, boxes, sacks and a frozen lamb carcass, hooves to the sky.

In Antarctica you do everything, and now I'm a butcher.

Roast pork loin first night—thanks, Marcos.

DECEMBER 23, 1995

WT   Pizza night
Custard with rum sauce and candied almonds
Cooking hell. Will I get organized?
Deliver cookies to Chileans

59°

DRAKE PASSAGE

SOUTH SHETLAND ISLANDS

62°

Ferraz
(Brazil)

Machu Picchu
(Peru)

Arctowski
(Poland)

Bellingshausen (Russia)

Fildes Peninsula

Frei/Marsh (Chile)

Artigas(Uruguay)

Stoney Bay

Flat Top Peninsula

Adm

Ardley Island

Great Wall (China)

King Sejong
(Korea)

Eco Nelson
(Czech Republic)

Jubany (Argentina)

Maxwell
Bay

NELSON ISLAND

59°

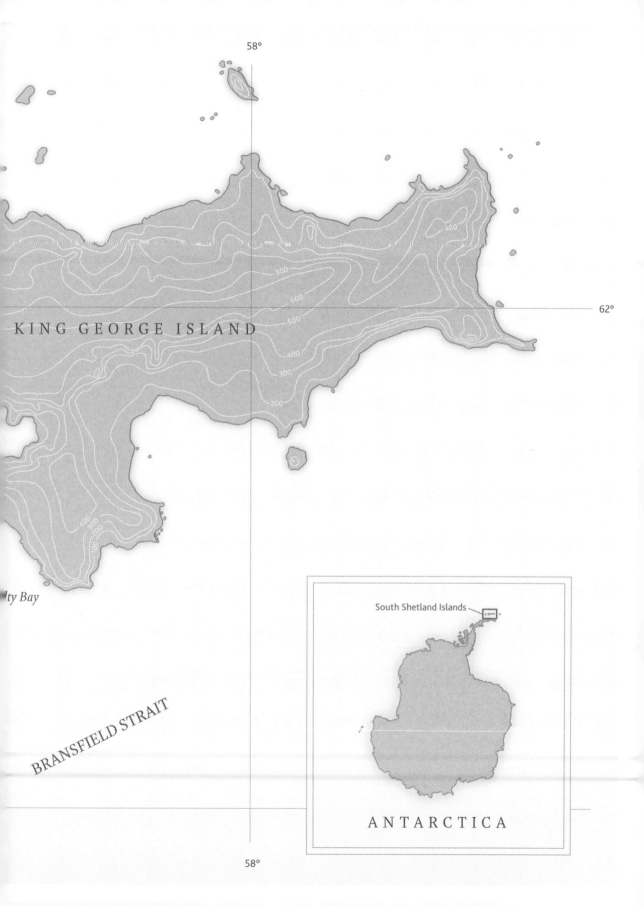

58°

62°

KING GEORGE ISLAND

ty Bay

400

500

600

500

400

300

200

400
300
200
100

BRANSFIELD STRAIT

South Shetland Islands

ANTARCTICA

58°

CD   Setting up the kitchen, common area and rooms for volunteers was exhausting. But now we've started the cleanup it's totally exhausting.

I'm glad we had a big breakfast. Pancakes. We were a work team of 12: five from VIEW Foundation in our blue coats and seven volunteers in parkas and hats. We met outside Canada House. Sky grey, sulky. It was chilly. We piled into facing seats in the PTS like a penguin huddle. Sasha the awesome silver-haired driver took us to Stoney Bay. The tank rolled up and down the hills. We couldn't see much outside the foggy little windows. At Stoney Bay, otherwise known as Fuel Tank Bay, were several tall fat rusted fuel tanks from the 60s. They were used to transfer diesel to ships. They sat on a long grey-black volcanic rock beach with small patches of snow. Behind us, hills. No penguins.

We surveyed the rubbish. Snakes of beige fabric fuel pipes with rusted nozzles, bits of metal, wire, plastic about. Got straight to work. Sean, Bob, John, Lou, Dewey, Dave and I heaved the first pipe across the beach, laying it out. The guys cut it with a stone-age looking saw. After we chopped the first pipe oil spilled out. Not cool. Well, it was not much. How old was the oil? John took a sample to check later. Sean will note how much spilled in the Environmental Impact Assessment. All of us including Sasha decided to wrap bundles of pipe in layers like puffed pastry dough. Took us about three hours per pipe.

I left the pipe team to check on Andrew, his mom and grandmom. They were hunched over collecting bits of metal and plastics and were pleased to show me they'd already filled the lower part of an empty gas barrel. How gorgeous this family joined us for the first camp and Christmas. No one was daunted by hard work or the drab weather. We kept pulling, cutting and bundling. It got chillier, rain almost turned to snow. Job won't be done today. I will be sore tomorrow. Hope everyone felt as good as I did. I'm sure they were just as tired and hungry. Loved Wendy's make-your-own pizza night.

Tonight I looked at a map in Sergey's office. Queen Maud Land, the Sabrina Coast, Marguerite Bay. So that's how women first got to Antarctica, without going. Their presence was missed after all. It's notable where you find women. Jeanne, the wife of early 20th century French explorer Charcot, apparently left him on the grounds of desertion.

DECEMBER 24, 1995
WT   Wind. Snow. Gift from Fernando the cook at Frei base—Pan de Navidad.

To do:
1. Grill eggplant, zucchini and peppers for Muffuletta
2. Move produce to storage shed, split boxes
3. Play with Cinnamon Bun recipe. Make fruit nut ring/wreath
4. Christmas presents with Lena

Lunch
· Rosemary Maple Borscht
· pizza leftovers or Muffuletta
· fruit & cookies

Dinner Menu
· Mulled Wine or Cola
de Mono*
· make your own fajitas: guac,
lettuce, tomatoes, peppers,
sour cream, cheese, onions,
meat, beans & rice
· birthday cake

Christmas Day Meal Prep
· make GORP
· stuff stockings with oranges
& candy
· gingerbread cookies
· Pan de Navidad
· Christmas wreath x2
· Russian Tea
· fruit salad with satin sauce
· soup
· bread & cheese

Christmas Dinner
· turkey with dressing
· roast daikon, yams
& onions
· mash with gravy
· cranberry sauce
· corn on the cob
· asparagus with dill butter
· green salad with croutons
· Bavarian apple torte

* *"Monkey's Tail"—traditional Chilean Christmas drink*

CD   We continued the pipe folding and bundling again at Stoney Bay. This is a much bigger job than one we can complete this summer. It'll take serious remediation and heavy logistics to remove the fuel tanks. But we're making a dent. Cold and damp yet great spirits. Volunteers who weren't on pipe duty used plastic bags for collecting general debris. I was impressed they filled three bags relatively quickly with small bits, wire, bolts, metal bits. Everyone seemed focused and it makes it worthwhile. Wendy is making a special dinner. Lunch was genius veggie borscht.

Tonight we were invited to the church service at Frei base, in a little blue wooden church on a rise. Five of our volunteers, four VIEW staff and Lena attended. I was asked to do a reading in Spanish. The sermon was on peace amongst people from different countries. After the service a distinguished gentleman approached me: the Chilean ambassador to the UN visiting Frei for Christmas. He wanted to hear more about our work.

Wendy gave Zuniga and his family Christmas cookie treats as a present.

With this first volunteer group we have three more days until we travel back to Argentina. I love this work I don't want to go back to the Foundation in Toronto but have to. I have other projects to plan and supervise: Costa Rica. Sean, Wendy, Lena and John will do a superb job. What a team considering how quickly we assembled.

I opened my present from my mom. Was hard to hold off. It was a small silver adult penguin and chick in an embrace. It made me teary. My mom and dad are supportive of all I do. I laughed though because while it was from my mom, the penguin caring for the chick was likely the father.

Fantastic dinner with the volunteers. We invited some guys over to sing. The only song the Russians knew in English was "Yellow Submarine." Bob said they're going to forever think the Beatles write Christmas carols. Sad to leave in a few days. The best Christmas ever.

Russian base: Bellingshausen, foreground (multi-coloured buildings), Chilean base: Presidente Eduardo Frei Montalva, background (red buildings), Collins Harbour on the Fildes Peninsula, King George Island on the Antarctic Peninsula. The 7 km long Fildes Peninsula at the southwest tip of the island is ice-free in summer. The island is 95 km long and 25 km wide at its most extreme points. 1996

Wendy, left. Carol, right, Bellingshausen, 1995

VIEW Foundation Camp 1 cleaning up abandoned fuel pipes at Stoney Bay, 1995

DECEMBER 25, 1995

WT   Cloud, strong wind, rain. Delivered wreath to Russians. Out of the deluge. All the volunteers are chipping in. Last night some helped knead dough and folk are taking to dish duty even though they're on holiday. Bob was on last night and chatted about long distance relationships—funny how the farther you go away from civilization the more you learn about people. Clever gifts from him at breakfast. All made from the dump. Mobile of cutlery and glass bits for me. For the boy in our group, a make-your-own penguin kit with bones from the beach. Great feeling of warmth and goodwill.

CD   I'm filthy, no time.

DECEMBER 26, 1995

CD   I don't want to leave. I said goodbye to Sergey. He said he was impressed with the work we did on Canada House and that it is the nicest place at the station. That Stoney Bay was already looking better. He promised everyone would be well taken care of after I left and all was set up. I look forward to following the adventure (and surely trials) I know the VIEW team will face. They're diversely skilled, competent and game.

I'm disappointed someone else from VIEW will go to help close camp, not me. I wish I could return for the final camp, the budget won't allow it and this is why we hired Sean. Better to be grateful for what I have than wish for more.

Bob's Xmas card note was sweet, "You have an inspired *view*. Keep on with your vision."

Lena came as I was shoving things into my bags. She took my hand and said, "This is a tradition before I leave a place. Sit on the bed for a moment." After a few seconds I jumped up. I saw my favourite toque under Wendy's bed. I hugged Lena. "You see, it is good to pause; it can change your perspective," she said.

WT   Volunteers changeover today. But Carol left and I'm quite sad. Spent time with Lena after dinner, conjugating verbs. First real sit-down Russian lesson and I feel better. Too tired to write. More later.

I work – Ja rabotaju
I go – Ja idu
I can – Ja mogu
Thanks, I can do it myself – Spasibo, ja mogu eta sdelat' sebya.

Chivalry not dead here.

Camp 1 Debris Collection (Dec. 22–26)
Stoney Bay Area 1: mixed waste & piping

DECEMBER 27, 1995

CD　I'm with the first volunteer cleanup group on the *Ioffe*. I'm seasick. It is awful. I suppose I'd like some company.

It was a sad goodbye yesterday to Wendy, Lena, Sean, John and the guys. I'm sure my seasickness has more to do with the Drake than our drinks last night.

Warm and lovely reception from staff and tourists on the *Multanovskiy*.

At dinner for some reason we talked Antarctic horror stories: the homicidal Uruguayan man they had to lock up. The Russian doctor who had an attack of appendicitis and decided to commit suicide but the other expedition members persuaded him they'd help him do the surgery on himself and assist with mirrors. Apparently he agreed.

Bob sent word for all the VIEW group and whoever else to meet at the bar. He worked hard at Bellingshausen and is the life of the party. His goal was to have the highest bar bill ever on the ship. He was treating. I wonder if we surpassed the record.

It didn't pay to be nervous about him coming on the project but I sincerely hope he found it useful, even if symbolically.

DECEMBER 28, 1995

WT　Camp 2. Odd group dynamics created by the nineteen-year-old kid. Never met anyone who could push my buttons this way. Never actually threatened ever, ever, ever, not to feed someone—not that he heard. Arriving fifteen minutes late to dinner was one thing, but ransacking the kitchen for ingredients to make himself an omelette when I had set aside dinner for him really set me off. I can see the others are weary of his antics. Gloria and Chizulko from Alaska coached me with understanding looks. They must have teenagers at home. Thought I'd made headway when he offered to do dishes until I found him up to his elbows in suds wearing my only set of baker's mitts. They'll take days to dry. Thank goodness for Lena. Laughed until we cried.

Things a little weird down in camp, too. At lunch Volodya Driver showed Sean my arm and asked "Pochemu?" ("Why?"), as if it's Sean's fault I'm covered in soot. And this afternoon a guy knocked on the sauna door as I was leaving and gave me a package claiming it was a Russian tradition: three pairs of bikini underwear with a colourful Spanish motif.

Lena seemed pleased to hear I didn't make a big deal about it and feigned disappointment when I told her. "It can't be much of a tradition; I didn't get any panties." Man, she's got good delivery. Helped diffuse the lingering creepiness of the way he repeated, "mirar, mirar," as if he wanted to see me in them. Checked my Spanish dictionary. Mirar: to look; to watch. Poor guy, he's got bush fever or the permafrost equivalent.

DECEMBER 29, 1995

WT　2:00 a.m. Carol, my friend, wish you were here and we could rehash the day.

Lena, Ilya, Radio Sasha, and Dima arrived 12:30 with Scotch to toast Sasha's birthday and I feel included. Ilya has wood-burned two signs for us on weathered planks: a Canada House sign for our door and one for my kitchen, Dietary and Aesthetic Laboratory. It's as if now that John and Sean are here it's okay to see Lena and me socially. Weird but good.

You'd be pleased with John and Sean; they're team players. Sean is pretty upbeat and has slid naturally into the camp manager role, but he could have waited a day or two before plastering all those site maps and work details everywhere, especially not in the dining room.

Stove still acting up. Kitchen walls are blackened, as is anything I bake or roast. Me too. No one knows for sure what is wrong, but we are dealing. Volodya Driver is making a valve in Diesel to control gas flow and Sean has found a passage in Lashly's diary that puts things in a poetic perspective. He had the bright idea of reading it at the beginning of each camp or on an as-needed basis—ours smokes when the wind is in the north.

> The ship is very comfortable, there is nothing whatever to grumble about as we live well, sleep warm and nice and have plenty of exercise. The only thing is we are troubled with a smoky stove when the wind is blowing hard. But I think other expeditions have suffered with the same complaint. He never got to the root of this trouble and when the wind was in the South the stove smoked so much that they had to do without it: Not nice in 52° of frost.
>
> —*Under Scott's Command: Lashly's Antarctic Diaries,* Discovery *Expedition 1901–1904*

Camp 2 Debris Collection (Dec. 26–30)
Stoney Bay Areas 2, 3 & 4: mixed waste
Bellingshausen Areas b1–b3: 1 barrel

> It's not so different from what I see at home: you know the bush, where they have done mining or forestry for a while then abandon the camps. It makes you look a little differently at what you've got in your own backyard.
>
> —Volunteer, anonymous

CD Still on the *Ioffe.* I'm feeling better. I gave a talk to the passengers about our first work camp at Bellingshausen and our aims. I miss everyone, especially Wendy and Lena.

Some grilling (why should we clean up after others) but good feedback.

DECEMBER 31, 1995
CD I spent a few hours in Buenos Aires on stopover with a wonderful man I met at the Ushuaia airport. Australian. Photographer. He was on another ship in Antarctica full of doctors studying penguins and the science of hypothermia. I think I'm in love. How is that possible? Details to follow. Everything since flying yesterday looks different.

Travel to Mongolia, study mosaic-making, write story.

What shall I do? It is my life and I can choose, which is the really exciting thing. 1996 looks great.

> Study the treasures of the Antarctic, even after I am dead.
>
> —Lt. Nobu Shirase, Japanese Antarctic Expedition, 1912

WENDY'S
ANTARCTIC
RECIPES

PITCH-AND-TOSS COOKERY

Wendy

———

I blame my mother for my cavalier approach to cooking. Always close by, she let me run free in the kitchen at an early age and whenever I'd ask how much of an ingredient to add or how long to cook something she'd reply, "about that much," "about that long," or "until it's done." It never occurred to me that there was a definitive answer or that I wouldn't be able to find one on my own.

That comfort with free-styling recipes played out in bush camps across Canada where I honed and refined my cooking style. A lack of ingredients was common in these circumstances, so when I couldn't follow a recipe to the letter I'd line up a number of cookbooks open to similar dishes and approximate a meal using available stores. Naming the final creations has always been part of the fun.

Conventional culinary wisdom maintains there is no room for guesstimates in baking. I would never advocate abandoning measuring ingredients, but I admit that when I make my chocolate chip cookies or honey oatmeal bread I never measure—except with my eyes and my fingertips. That said, these recipes have been painstakingly developed, especially for baking. I want you to be successful when you try them at home so I used best-practice rules when measuring flour—I spoon it into a dry measuring cup and level it off using a knife. I don't do a lot of sifting when I bake, but I would never forgive myself if someone bit into a baking soda deposit in a cookie or muffin. After measuring I put the baking soda in the palm of my hand and press out any lumps. Whisking dry ingredients together in a bowl helps as well.

You'll find these recipes are written in an idiosyncratic style, but I hope they'll encourage you to taste, look for visual clues and develop a real feel for whatever you are preparing. You'll be pitching and tossing before long.

HONEY OATMEAL BREAD (WENDY BREAD)

Of all the types of bread I bake, my honey oatmeal bread is the recipe I fall back on most often. Its modest beginnings are in porridge left over from breakfast. With a generous amount of honey in the mix it bakes into soft lovely loaves that disappear quickly. I typically make a much larger batch than this so that I can turn out some of the other recipes I use this dough for at the same time, but I've scaled back the ingredients to create this master recipe. It translates beautifully into Fruit Nut Rings (page 95), Cinnamon Buns (page 91), or Pizza bases (page 147).

4 cups water // 2 cups large-flake rolled oats // 1⅓ cups skim milk powder // ¼ cup butter // 1 tablespoon salt // 1 cup honey // 1 cup warm water // 1 teaspoon sugar // 2 tablespoons active dry yeast // 10 to 11 cups all-purpose flour

Bring the 4 cups of water to a rolling boil. Stir in the oats and reduce the heat to medium-low. Cook the oats, stirring often, for about 10 minutes to make a thick and creamy porridge.

While the porridge is cooking prepare all the other ingredients. Put the milk powder, butter, salt and honey into an oversized bowl. (Try a salad bowl or wash basin if you don't have a mixing bowl large enough.) Have the flour measured and set aside in bowls according to the increments it will be added: one 7-cup and four 1-cup measures. Finish your *mise en place* by whisking the sugar, yeast and one cup warm water together in a small bowl. Set aside until frothy.

When the porridge is ready add it to the large bowl and whisk vigorously until the butter has melted and any clumps of milk powder are evenly distributed. Squish any stubborn clumps against the side of the bowl. Let cool to lukewarm.

By this time the yeast should be frothy. Add it to the porridge mixture followed by the 7-cup portion of flour, mixing well. Gradually stir in additional flour until you are no longer able to use a spoon, then dust your hands and scrape the dough from the bowl to form a sticky ball.

Turn the dough out onto a floured surface and knead, adding as much of the remaining flour as is necessary to keep it from sticking. At times the dough may appear to resist the flour, but it may still be quite soft on the inside. If you cut the dough into two equal size parts at this point it is easier to gauge how much more flour is needed. I often cut, check for stickiness, add flour and reunite the dough as many as three times during my kneading routine. I know it's ready when the dough resists slightly when I push it away, but then snaps back without leaving any lingering stickiness on my hands. This will take 10–15 minutes—the longer the better. Depending upon the weather, your location and the time of year you could have as much as one cup of flour left over.

Return the dough to the bowl after greasing it with butter, turning to coat the dough all over. Cover with a tea towel and put it in a warm place to rise for about 1½ hours.

To make the bread

When the dough has doubled in size turn it out onto the floured surface and bring it down with a punch. Knead four or five turns then cover and let it rest for a couple of minutes while you prepare the baking sheets or bread pans.

I'm fond of making three-strand braids with this dough because they bake so quickly and encourage a delicious informality wherever I serve them. It makes equally lovely conventional loaves, free-form boules or dinner rolls. To make conventional loaves follow the shaping directions for Cook's Bread on page 59.

For braids, divide the dough into three equal parts. Knead each one into a uniform ball and set aside. If you have three baking sheets you can work through all the dough at once and control the rise of the braids by keeping them cool. Otherwise the balls of dough will have to wait their turn in the refrigerator and be formed as pans become available.

To form the braids, use a rolling pin or your hands to press each ball into a 6 by 12-inch rectangle. Cut the dough into three identical strands and then roll each of them into uniform ropes of equal length, about 24 inches long. A good technique is to begin with your hands together in the centre of the strand and gradually move them outwards to the ends as you roll. Place the ropes side by side to complete the braids. I find if I begin to braid in the middle and criss-cross outwards, one direction at a time, I can avoid stretching or breaking the strands. Join the strands together with a pinch and a tuck at both ends. Repeat for the remaining balls of dough, transferring each of the finished braids to greased baking sheets as you go.

Cover with a clean tea towel and set aside to rise for about one hour or until doubled in size. Bake in the centre of 350°F oven one pan at a time until golden brown, 20–25 minutes. Remove from pan immediately and let cool on rack. Served warm, it pulls apart beautifully. Perfect for dipping.

For smaller braids divide the dough into six parts and press each of them into a 4 by 9-inch rectangle. After you make the strands you'll want to roll each one into a 15-inch rope. Continue as above but with a baking time of 10–15 minutes.

Makes three large or six small braids.

This bread freezes well either baked or unbaked. To freeze unbaked braids, place them on a baking sheet lined with wax paper, cover tightly with plastic wrap and put the pan in the freezer. After the dough has frozen solid, wrap each braid entirely with plastic wrap, return it to the freezer and reclaim your pan. Alternatively you could use a cardboard flat for canned drinks lined with waxed paper. When you are ready to bake, transfer the braid to a prepared pan, cover with a tea towel and let it thaw and rise until doubled in size, 6–8 hours. I like to take it out of the freezer in the morning to bake it in late afternoon.

ROASTED GARLIC IN HERBED OIL

The inspiration for this dish was a four-foot-long double braid of Argentinean garlic that hung on the wall of my Antarctic kitchen. It is really more method than recipe, as much depends upon mood, pan size and the herbs you have on hand. These proportions fit nicely into a small cast iron skillet— feel free to ad-lib with the herbs. I use fresh instead of dried when available and once on a whim in Antarctica I added a handful of the cranberries the Russians had given me. I like to keep them on hand at home now for the vibrant burst of colour and unexpected flavour and texture they bring to this dish.

6 to 7 heads of garlic // olive oil // 1 teaspoon coarse salt //
½ teaspoon each of peppercorns (try pink or green if available),
chili flakes, dried rosemary, sage and thyme // a few bay leaves //
a handful of frozen cranberries

Preheat the oven to 350°F. Cut ¼-inch from the tops of the garlic heads to expose the cloves. Remove any messy bits of skin, but not so much as to compromise the bulbs' structural integrity.

Place the garlic in a skillet or ovenproof dish. Five to six arranged around the perimeter and one in the middle works well—leave a little space in between. Add enough olive oil to come a ¼-inch up the side of the pan and sprinkle with salt, peppercorns and chili flakes. Turn the bulbs to coat with oil and reposition them cut side up. To finish, tuck as many bay leaves as you think looks nice between the bulbs and then scatter the remaining herbs over top, crushing them between your fingers as you go to help release the flavours.

Cover tightly with foil and bake until the skins on the garlic bulbs are golden brown and the cloves are tender. This should take about 50 minutes, but check after 40 minutes and remove the foil to hasten browning if necessary. Add the cranberries 5 minutes before removing the baked garlic from the oven. You can roast the garlic ahead of time and add the cranberries when you are reheating.

Serve tucked in beside warm braids of Honey Oatmeal Bread. I like to encourage the use of hands to pull away and squeeze the cloves onto the bread, but I always have two small spoons at the ready for the less adventurous. A dish for the skins is a good idea.

Makes enough for twenty.

MULLED WINE

I love the look on people's faces when they catch the first whiff of mulled wine simmering on the back of the stove. The sweet, heady blend of spices lifts even the most mediocre vintage, and the high ratio of orange juice to wine is an advantage if you are trying to eke out dwindling provisions.

10 large oranges (9 for juice and 1 for the pot) // 1 teaspoon whole cloves //
4 allspice berries // 4 cinnamon sticks // 5 cardamom pods // ¼ cup brown sugar //
1 bottle dry red wine (750-ml)

Place a sieve over a large pot. Roll nine of the oranges to loosen their juices, then cut and ream each half over the sieve into the pot. Place the pot over medium-high heat.

Slice the remaining orange; spike the slices decoratively with cloves and add them to the pot along with the allspice berries. Next break the cinnamon sticks and crack the cardamom pods; add them to the pot. When the juice begins to steam and bubbles start to form around the side of the pot stir in the brown sugar and reduce the heat to medium-low. After about 10 minutes pour in the wine, cover, and reduce the heat so that it is just steaming.

How long you mull the wine is a matter of personal preference. I don't obsess about how much alcohol might burn off and simmer it gently for another 40 minutes so that all the flavours get to know one another. And I admit I don't pay much attention to how long it has been on the back of the stove as guests begin to fill the kitchen over the course of a gathering. Cut it with more orange juice or top it up with wine if you need to bring it back to life and think of the syrupy spicy goodness left at the bottom of the pot as starter for your next batch.

Makes enough for six to eight people.

ROSEMARY MAPLE BORSCHT

Vladimir the Russian cook made his borscht using a meat stock. My version kept the vegetarian volunteers in camp happy and even got the thumbs-up from the Russians. To make vegan Rosemary Maple Borscht just substitute olive oil for butter and hold back on the dollop of crème fraîche or sour cream.

2 pounds beets (around 5 medium) // 3 medium potatoes // 2 tablespoons butter // olive oil // 2 onions // 2 cloves of garlic // 1 celery stalk // 2 large carrots // 1 small cabbage (about 5 cups chopped) // 1 tablespoon caraway seeds // 8 cups water // 3 tablespoons apple cider vinegar // 3 tablespoons maple syrup // 1 can crushed tomatoes (28 ounces) // 1 tablespoon sea salt // pepper // fresh rosemary

Peel and cube the beets and potatoes and put them aside. Heat the butter in a large pot set over medium heat and add the beets and potatoes, tossing to coat them with butter. Reduce the heat and sauté, stirring occasionally with a wooden spoon and being careful not to bruise or break the cubes. After about 5 minutes add enough water to cover the vegetables and gently simmer until tender, around 10 minutes.

While the beets and potatoes are cooking, mince the garlic and onions and chop the remaining vegetables. Put the caraway seeds into a large Dutch oven or stock pot and toast them over low heat, pushing them around the pan from time to time so they don't burn. When you begin to smell the aroma of the caraway add enough olive oil to generously coat the bottom of the pot. Stir in the onions, garlic and celery, sprinkle with salt and cook over medium heat until the vegetables are soft and translucent. Next mix in the carrots and cabbage and sauté for about 5 minutes before adding the remaining water. Bring briefly to a boil and reduce the heat before making the final additions.

Add the beets and potatoes in their cooking liquid, along with the vinegar, maple syrup, crushed tomatoes and a large sprig of fresh rosemary. Cover and simmer for at least 40 minutes to bring the flavours together. Season to taste and make adjustments to the thickness of the soup by adding water as you see fit. Garnish with rosemary and a dollop of crème fraîche or sour cream and serve with freshly baked bread.

Makes enough for ten to twelve people.

Cooking for small teams of volunteers on King George Island meant I had to scale back my recipes from my bush cook days, but only so far. I love that I can get a few meals from this soup. It keeps for five days and freezes well even if you aren't in Antarctica.

CINNAMON BUNS

*I made my first Cinnamon Buns in my early teens on a canoe trip using a prepared biscuit mix
and the bottom side of a canoe as my work surface. I switched to a yeast dough not long after I learned
how to make bread and have been tweaking my approach ever since. My breakthrough moment
came when I started to add oats to the filling and topping. There is a decent argument for the nutri-
tional value they add, but I think their caramelized goodness is more compelling.*

1 batch Honey Oatmeal Bread dough made through the first rising (page 81) //
1½ pounds butter at room temperature // 5½ cups packed brown sugar //
¼ cup cinnamon // ¼ cup poppy seeds (optional) // 2 cups large-flake rolled oats //
½ cup all-purpose flour

Set aside ¼ cup of the oats in a medium bowl. Cream the butter and brown sugar together in
a large bowl. Stir in the cinnamon, 1¾ cups of the oats and the poppy seeds if using.

Using a dry measure, add 1¼ cups of this mixture to the reserved oats and set aside to make
a streusel topping later.

When the dough has doubled in size turn it out onto a lightly floured surface, punch it down
and cut it into four equal parts. Knead each piece a few turns, roll into uniform balls and set
aside to rest while you grease your pans.

As this is a high-yield recipe you'll likely need to use a variety of pans—when making your
selections look for pans with high enough edges to support the rolls. I typically use a combination of baking sheets (with rims), cake pans and pie plates with the intention of keeping
some cinnamon buns for home, sharing with friends and freezing a quantity. Heading into
busy social seasons I find it really handy to have a stash of hostess gifts at the ready. They
freeze well either baked or unbaked.

To form the cinnamon buns, roll out each ball into a 7 by 21-inch rectangle. Spread a quarter
of the filling (1¾ cups) evenly over the dough, pressing it in as you go. Starting at the long side,
roll up the dough like a jelly roll. Pinch the seam to seal together and gently roll the dough
to make it more uniform.

Using a sharp knife cut the roll into 1-inch slices. Use your hands to coax each piece back into
a circle and place it onto a prepared pan. I like to nestle them closely together (up to ¼-inch
apart) so they don't unravel. If you are using a pan with shallow sides you will want to leave
space around the perimeter; otherwise you may make a mess of your oven when filling spills
over the sides. Repeat with the remaining portions of dough and filling.

Cover each pan with a tea towel and let rise in a warm place until doubled in bulk, 45 minutes to an hour. If your kitchen is really warm you'll want to rotate some of your pans into a cooler spot to control the rise. I like to let them rise in the refrigerator overnight to bake for breakfast. An overnight rise also works well for frozen dough; take the rolls out of the freezer before bed and let them thaw and rise on your counter.

About 15 minutes before baking, preheat the oven to 350°F and add $\frac{1}{2}$ cup flour to the mixture reserved for the streusel topping. Work it together with your fingers and then top each roll with a $\frac{1}{2}$- to 1-teaspoon-sized knob, depending on the size of the rolls. Some topping will ooze off the sides as the buns bake. Ideally you want to keep most of it mounded on top.

Bake in the centre of the oven until golden brown, 25–30 minutes.

Makes $5\frac{1}{2}$ dozen cinnamon buns.

Scaled-back ingredients to yield $2\frac{1}{2}$ dozen cinnamon buns:

For the dough
2 cups water // 1 cup large-flake rolled oats // $\frac{2}{3}$ cup skim milk powder //
2 tablespoons butter // $1\frac{1}{2}$ teaspoons salt // $\frac{1}{2}$ cup honey // 5 to 6 cups all-purpose flour //
$\frac{1}{2}$ cup warm water // $\frac{1}{2}$ teaspoon sugar // 1 tablespoon active dry yeast //

For the filling and topping
$\frac{3}{4}$ pound butter at room temperature // $2\frac{3}{4}$ cups packed brown sugar //
2 tablespoons cinnamon // 2 tablespoons poppy seeds (optional) //
1 cup large flake rolled oats // $\frac{1}{4}$ cup all-purpose flour

FRUIT NUT RINGS

My Cinnamon Buns dressed up with fruit and nuts then baked in the shape of a wreath. This recipe yields two rings so I like to play around with the way I decorate the tops. Our Christmas morning wreath in Canada House and the one we shared with the rest of the camp were both decorated with dough cut-outs representing each of the volunteers, the VIEW team and the Russians. A simpler version is to score the ring with decorative slashes and sprinkle it with coarse sugar. A spectacular variation worth trying is a Fruit Nut Twist.

1 batch Honey Oatmeal Bread dough made through the first rising (page 81) //
$\frac{3}{4}$ pound butter at room temperature // 2$\frac{3}{4}$ cups packed brown sugar //
2 tablespoons cinnamon // 1 teaspoon nutmeg // 1 teaspoon vanilla // $\frac{1}{3}$ cup currants //
1 small orange // 1 cup dried apricots // $\frac{1}{2}$ cup walnut pieces //
$\frac{2}{3}$ cup large-flake rolled oats // 2 tablespoons poppy seeds // 1 egg // 1 teaspoon water //
2 teaspoons coarse sugar

Cream the butter and brown sugar in a large bowl then stir in the cinnamon, nutmeg and vanilla. Put the currants in a small bowl, cover with warm water and set them aside to plump. Measure and prepare all the other ingredients: juice and zest the orange; chop the apricots; drain the currants. Add them to the butter and sugar, mix well, then stir in the walnuts, oats and poppy seeds. Set the filling aside until the dough has doubled in size.

When the dough has completed its first rise turn it out onto a lightly floured surface, punch it down and divide it into two equal portions. Knead each piece a few turns, shape into uniform balls and set aside to rest, covered, while you grease your pans. A 10-inch springform, a Bundt pan or an angel food cake pan all work nicely.

To form your rings, roll out a ball of dough into a large rectangle, about 20 by 11 inches, on a floured surface. If you are planning to decorate the top with cut-outs you should start with a larger rectangle and set aside the scraps of dough to use later. Spread half of the filling (around 2$\frac{1}{2}$ cups) over the dough, pressing it in as you go.

Starting on the long side, roll up the dough like a jelly roll. Pinch the seam to seal together and gently roll the dough to make it more uniform. Place seam side down in a prepared pan and pinch the ends together to seal. Cover with a tea towel and let rise in a warm place until doubled in size, 45 minutes to an hour.

I like to make the second ring right away. If your kitchen is warm and you find the second ring is rising too quickly, put it into the fridge and take it out just as the first ring is going into the oven.

Fruit Nut Twist

To decorate the ring with cut-outs, roll out the scraps you've reserved about 20 minutes before baking. Cut into desired shapes using a knife or cookie cutters. Moisten the back of each cut-out with a little water and gently press onto the ring.

About 15 minutes before baking, preheat the oven to 350°F and whisk 1 egg with 1 tablespoon of water to make a glaze. Gently brush it over the rings.

For decorative slashes use a very sharp blade to score the ring just before baking. Sprinkle with coarse sugar after you have brushed on the glaze.

Bake until golden brown, 55–65 minutes. If the cut-outs are browning too quickly cover with a bit of foil. Run a knife around the edge of the pans as the rings cool on racks.

Makes two rings, each yielding about 16 slices.

To make a Fruit Nut Twist

Place the roll seam side down on your work surface and cut it in half lengthwise. Keeping the cut sides up, twist the two strands of dough together and place into the prepared pan. Pinch the ends together to seal. Cover with a clean tea towel and set aside to rise. Follow the baking instruction as above.

"Many sweet memories crowd on me as I lay in my bag meditating the last day of the year. Home & faces and places & of our present position that one cannot altogether regard as sweet. Drifting about on an ice floe & 18.9 miles from nearest known land. Still, to apply our old sledging motto, "It might be much worse." We have plenty of good grub & with the coming warm season & subsequent dissipation of the ice are enabled to greet with cheery aspects the New Year 1916."

—Frank Hurley, December 31, 1915

A radiant turret lit by the midnight midsummer sun, Frank Hurley. 1911–1914

CLEANING A CONTINENT
Carol

———

> If you want to inspire young people to preserve Antarctica thirty years from now,
> you've got to have a great story to show that you've already been doing that.

—Sir Robert Swan, 2012

By 1995 everyone knew earth had a garbage problem. In Antarctica waste doesn't decompose. The Antarctic Treaty devoted the continent to peace and science. It also banned nuclear testing and radioactive material dumping. The Treaty's additional Madrid Protocol (entered into force in 1998) set standards for human activity, instructing nations to manage waste, protect the environment and conserve flora and fauna.

Lena, our Russian liaison officer at Bellingshausen had previously worked at a cleanup at Novolazarevskaya station in Queen Maud Land.

We had no idea another cleanup was going to happen or if it was already being planned when I visited the Russians in St. Petersburg to plan the VIEW Foundation project. However several months after we departed Bellingshausen, polar explorer Sir Robert Swan took 35 young volunteers from 25 nations to start what became a long-term cleanup of Bellingshausen with the Russians.

Swan returned with youth groups from 1996 until 2000 to clean and in 2002 removed 1,500 tons of solid waste and rubbish for recycling in Uruguay. This was the largest Antarctic cleanup ever. Swan also helped develop a waste management plan and a renewable-energy-powered education facility and continues collaborative cleanup efforts today.

The 30 or so countries with Antarctic bases report their environmental activities including remediation, debris removal and water and waste management. Independent groups and international inspection teams monitor compliance with the Antarctic Treaty, including that of tourist operations. Despite progress and innovations, some countries still must do more to comply.

Russia highlighted their cleanup efforts at the Antarctic Treaty Consultative Meeting in 2002. "The first experience of cooperation with non-government organization was acquired in 1995–96 summer season when the Canadian ecological View Foundation worked at Bellingshausen...As a result 20 fuel drums of small waste and 500 meters of old fuel hoses were removed out of Antarctica."

Resource extraction in Antarctica might be difficult logistically and politically but it's not unthinkable that countries, including Russia, might try. Hopefully by 2041 sustainable technologies will replace dependency on hydrocarbons and we will have long realized the urgency to change our destructive footprint on the planet. During our time with our Russian colleagues and the volunteers, it was crystal clear that human relations were invaluable when trying to turn words into action.

SIXTEEN
GRANDFATHERS

JANUARY 2–15, 1996

Tonight the east wind is blowing up a winter-like summer storm.

JANUARY 2, 1996

WT   Camp 3. Only two volunteers this time and as they say, less is more. I really like these two, but poor Lena, Sean and John are struggling to keep them motivated in this constant wind and drizzle. It's hard to build a team when staff outnumbers the volunteers.

Highlights: Much excitement leading up to New Year's. Running errands anywhere in camp an occupational hazard as holiday rations have been distributed. Coaxed—no pulled—in for "chut" "chut" (a wee bit of) vodka, chocolate and fruit with the camp elders as I passed by Sergey's office, and then again at Diesel. Yummy Diesel room fish—Notothenia (rock cod).

New Year's Eve potluck with the Russians. I brought Mexican and served mai tais with glass tubes Dima found in the lab to use as straws. Enduring images of the guys sipping from fresh pineapples and struggling to construct their fajitas as I coached them. I'm becoming even more attached to these men. Feel so fortunate to be part of it all. Raucous singing throughout the night almost as if they'd forgotten Lena and I were there. John and I guess Sean think otherwise. I feel like I'm holding my own so I don't know what to make of the suggestion that the two of them take turns sleeping in Canada House to ward off "predators." What is this, *The Bodyguard*? Ah well—it comes from the right place I guess.

Delivered hot drinks and fondue to the team at Stoney today. It takes some planning, but it's great to get out of Canada House and go for a hike. After lunch John and I took turns with the hip waders to gather kelp floating in the bay. Made Sea Cabbage Salad for dinner—we agree we wouldn't want to live on it.

The smell of frying greeted us on awakening this morning, and at breakfast each of us had two of our nutty little Notothenia fish after a bowl of porridge. These little fish have an extraordinarily sweet taste—bread and butter and marmalade finished the meal.

—Robert Falcon Scott, *Scott's Last Expedition*, 1911

JANUARY 3 , 1996

Changeover day. Two new volunteers.

Speedo-clad scientist emerging from Maxwell Bay between the Zodiacs pulled up on the beach: "You and me sauna next Saturday?"

Me (loud enough to get a laugh from one of the outgoing volunteers): "In your dreams."

Camp 3 Debris Collection (Dec. 30–Jan. 3)
Stoney Bay Areas 5–7: mixed waste

JANUARY 6, 1996

Sasha finally got a call through to P for me. Not like P to hold out on a "ti amo" sign off, but maybe the radiophone threw him off. Smarts a bit.

Got through to the VIEW office as well. Don't know how Sasha is doing it without Inmarsat satellite hook-up, but relieved I didn't have to pay $7/minute for the calls. Carol says the 80-year-old woman who walks with a cane has cancelled. And the uncertainty over where the garbage will go has been resolved. It makes sense that it goes back to Russia, but it would be simpler if it could go to Ushuaia. I wonder how much the garbage disposal fine is.

Great hike with John for a picnic near Flat Top this afternoon. His love for this place and the Russians is palpable. Poor guy—heading back to Georgia tomorrow.

Camp 4 Debris Collection (Jan. 3–7)
Stoney Bay Area 8
Bellingshausen Areas 1 and 2: 2 barrels mixed waste

Experiencing daily life on the windy continent makes for a better conception of the rigors of living in a harsh climate. The best part of the trip was the first hand look at the effects of time on man-made materials—picking up trash totally unchanged i.e. the plastics and cloth, copper covered wire, ionized metals.

—Volunteer Carolyn, Washington

JANUARY 9, 1996

11:50 p.m. Tonight the east wind is blowing up a winter-like summer storm and we are experiencing the first real accumulation of snow. Canada House is snow-packed and the whole place rather magic. Lying in bed listening to the wind howl. End to the perfect day.

Breakfast in bed. 9:45 a.m. meeting at Diesel to go fishing with Volodya Driver. Warnings from all the guys assembled for the send-off: wear a warmer hat; take sunglasses; wear mittens—and so on and so on. Life with sixteen grandmothers.

Sergey, Vadim and Doc Sasha in Sergey's office, Bellingshausen, 1995

Canada House and storage shed in background, Bellingshausen, 1995

Drive to Stoney Bay. Volodya is gallantly determined to lift me over a trickle of water and up a small rise on our way to the fishing hole. It's a harmless dance, so I let him. Dropped our lines between the crags and pulled in six in no time. Trusler two, Volodya four and then off to the shelter to warm up. I love those sheds, especially the wooden one, but he's rigged the mini Quonset with gas. Once inside he takes my socks from me to squeeze out the water and hold them up to the flame. So intimate I had to look away and try hard not to spoil the perfectness of the place with the please-don't-think-this-is-a-date soundtrack ringing in my head.

Back to Diesel for tea with the guys on duty. Volodya tells everyone assembled I caught all six fish. I'd love to draw or photograph this room, but how to capture the mood without spoiling it, changing everything? Lunch arrives. Encouraged to eat to fatten up and keep warm. Meat patties and raw onions on Cook's Bread with an orange for dessert while my socks dry. Again with the socks.

Conversation with Volodya afterwards which I understand as follows: he loves to do things for me, he's separated from his wife since 1988, he has no home to go to, and then something about fifteen years, which I think is our age difference. Now that and his invitation to go fishing again, I think I should leave alone. That and the comment about my eyes being the colour of tea.

Quiet dinner with Lena and Sean. Our first with only the three of us. Very civilized with candles, fish, fried rice and wine. Tolya is definitely grandmaster of the fish fry. I must learn from him. Down to the mess after dinner. Guys out for a smoke on the stoop. Couldn't resist staggering, swinging the vodka bottle as I approached. Had them going for a minute. "Problema Vendi?" They laughed when they saw the bottle was full of pepper for Volodya Cook. Then videos with the radio hut guys—a lot of shuffling and throat clearing during the love scenes.

And now to sleep.

JANUARY 11, 1996
Dreams to achieve:
· letters
· review old menu, make new extended menu
· journal
· language
· sort food in shed
· helicopter excursion with Uruguayans

JANUARY 14, 1996
· White Bean and Roasted Garlic Pâté with Focaccia
· Roast Beef with Yorkshire Pudding and Gravy
· Roasted Carrots
· Baked Stuffed Onions
· Custard with Fruit Compote

Tourists from the *Aurora Australis* icebreaker next to wedge iceberg, Antarctic Peninsula, 1995

JANUARY 15, 1996

A bit of a disaster this camp—and a thorn in our sides, but one to learn from. We've been upfront about our limitations and make do when the program falls short of trip literature promises but how to win over those who don't see value in the contribution we are trying to make. People are funny. Struggle to overcome this, especially for Sean and Lena. Hope I didn't offend when I shut down the volunteers grumbling about missing tossed salads.

Camp 6 Debris Collection (Jan. 11–15)
Stoney Bay Area 9: mixed waste
Bellingshausen Area 3: 1 barrel mixed

———

Every project needs a naysayer—To be fair, I think everyone went through it to some degree. After ten days sailing pristine Antarctic waters, no travel brochure could ever prepare you for Belling-shausen beach and its squalor. Often I'd read disappointment on faces as early as dinner the first night. Or it might come out at breakfast, after a bad sleep on a lumpy cot or after experiencing a foul communal WC when accustomed to ensuites. By lunch people might be talking it through with good humour or silently stewing. A chance to live and work in Antarctica—a trip of a lifetime—wasn't meeting expectations or adding up to the $4,000 price tag.

We sailed through the first part of the project. Even if volunteers weren't directly involved with the heavy work, removing defunct pipe and hoses from the fuel depot at Stoney was universally satisfying. We made visible progress every day. A Uruguayan pilot buoyed spirits further when he told us he'd noticed a difference from when he was last stationed on King George Island in 1994.

But in January as snow melted and exposed new sites, the focus of the cleanup shifted to include more tedious work around camp and an old dump. Squatting for hours on end, filling a kitchen bucket piece by piece with small bits of glass, metal and wood wasn't as tangible and certainly not as pleasant as the teamwork at Stoney Bay.

Factor into the equation a projector that didn't survive the Drake Passage, no VCR for the advertised "formal lectures", no motor for the Zodiac for "sunset tours around icebergs," no sunsets in December that far south and you could understand why patience might dwindle. Well, yes and no.

Everyone has his or her own arc. Perhaps we should have sent a thank-you note to that disgruntled pair in January who balked at picking up someone else's trash and made pejorative remarks about the Russians on their project evaluations. Sean rose to it. He increased interaction with the Russians, organized soccer games with the other bases—even the dinner parties were his idea and he had the personality to pull them off. Long walks with or without a "naturalist," getting to know the neighbours, and lingering around the dinner table all have value. On the work front, from that time on, he and Lena encouraged volunteers to articulate their expectations, made meeting them with grace and tolerance a daily focus and stuck to task no matter what anyone said.

As for my snarky salad comment I'm pretty sure I simply explained in the nicest possible way why fresh baby greens weren't feasible and wondered if they would have liked Sea Cabbage Salad.

SPICED TEA (RUSSIAN TEA)

I brewed this tea for hundreds of tree-planters and served it for years without knowing what made it Russian. My starting point was the tea a friend's mother made by blending powdered tea, juice crystals, spices and sugar. She called it Russian Tea, so I held on to the name when I adapted her recipe to take the sting out of cold mornings in tree-planting camps. After my first taste of the sublime brew Sasha poured for me in Diesel I felt self-conscious calling mine anything more than Spiced Tea.

To make Sasha's Russian Tea, pour hot water over a handful of loose black tea leaves nestled in a small strainer set over a glass. Squeeze the juice from a large wedge of lemon and add it to the glass with plenty of sugar. Stir. It tastes better if you take time to savour it, especially with company. My Spiced Tea recipe is better for a crowd.

3 cinnamon sticks // 2 tablespoons whole cloves // water // 1 cup sugar //
4 oranges // 4 lemons // 3 tablespoons loose black tea

Break the cinnamon sticks into a mid-sized pot containing 5 cups of water and set over medium-high heat. Zest one of the oranges and add it to the pot along with the cloves. Sprinkle in the sugar, stirring to dissolve as the water is coming up to temperature. Bring it briefly to a boil, then reduce to medium-low and simmer until the aroma of the spices begins to fill the room.

Bring 8 cups water to a gentle boil in a large pot.

While the syrup is simmering, ream the remaining citrus into a small bowl. Remove the syrup from the heat after about 10 minutes and add the juice, pouring it through a fine sieve. Add the tea leaves, then the 8 cups of boiling water and steep for at least 5 minutes. Strain the hot tea into a teapot or thermos, or directly into individual cups or glasses.

Spiced Tea keeps in the refrigerator for up to a week and is refreshing hot or cold. Flavours will intensify as the tea matures—at any point you can adjust the strength by adding more water or arrest the process entirely by straining the remaining brew.

Makes enough for twelve to fourteen people.

CHEESE FONDUE

I laughed when one volunteer with her mouth half-full of bread and cheese suggested, "If Wendy had been along Scott would have made it." But she might have been onto something, at least with respect to fondue. High in both fat and protein, with the added benefit of creating a sound architecture for camaraderie, the combination of molten Emmentaler and Gruyère could be the perfect expedition repast—that is if you overcome the inconvenience of hauling crates of wine.

½ pound each of Emmentaler and Gruyère cheese //
3 tablespoons all-purpose flour or 1½ tablespoons cornstarch // 2 cloves of garlic //
2 cups dry white wine // juice of ½ a lemon // nutmeg // pepper

Prepare the cheese by dicing or cutting it into strips and dredge it lightly in the flour or cornstarch. Using flour makes a robust, opaque fondue while cornstarch turns out a marginally lighter version that also happens to be gluten-free. They are equally delicious.

Peel the garlic cloves, smash with the flat side of a knife and rub the pieces around the inside of a four-quart pot. Some people like to discard the garlic at this point, but I think adding it to the pot gives the fondue a bit of kick and provides and element of surprise for the lucky diner who finds it at the end of his or her fork.

Pour in the wine and bring it to a gentle simmer over medium heat. When bubbles begin to rise to the surface add the lemon juice and the first handful of cheese, stirring constantly with a wooden spoon. Bit by bit, as each addition of cheese begins to melt, slowly add more, stirring in a figure eight pattern to prevent the fondue from seizing. Once the mixture is smooth, stir in a pinch of nutmeg if desired, and a good amount of freshly ground pepper.

Set over a low flame to keep warm and serve with cubes of French or Italian bread. Round out the meal with cold meats, sausage, pickled onions and gherkins. Wedges of apples or pears and steamed vegetables such as broccoli, green beans or asparagus are worthy alternatives to dipping bread.

Makes enough for a meal for four—more if you are serving it as an hors d'œuvres or starter.

FISHERMAN'S FISH

I don't know what the guys from Diesel did during their downtime throughout the rest of the year, but it seemed to me as soon as the temperature pushed above 0°C they started jigging for Notothenia or rock cod in Stoney Bay. It was mostly the guys who were on call for long stretches who liked to fish—the petroleum engineers, the mechanics; the guys who kept the base running. After completing an overnight shift they were the only ones with time off during the day.

Tolya made the best, it was universally understood—that much I picked up the first time I was invited for fish in the systems control room at the power station. Sasha was on duty that early January day. He and Tolya had jerry-rigged a hot plate and plugged it into one of the circuit-breaker panels lining the walls. Tolya cut the fish crosswise into small steaks—almost like chevrons. He dusted each piece with flour, salted and then fried it to crispy perfection. Vasiliy, Sergey and Fernando, the cook from Frei Base, arrived just as the first batch was being served.

No dishes. No forks. You eat Fisherman's Fish with your hands using your fingers to pull the tender flesh away from the bone. I make it at home using the whitefishes our local fishmonger brings in. Freshly caught bass, trout, pickerel or perch would be even more delectable.

2 whole fish about 1 pound each (whitefish, cleaned, with the skin on) //
$\frac{1}{4}$ cup all-purpose flour // coarse salt // vegetable oil

Cut the fish into $\frac{1}{2}$- to $\frac{3}{4}$-inch steaks and pat dry. Put the flour on a shallow plate and sprinkle with salt—a few pinches should do. Add enough oil to a large skillet to cover the bottom and place it over medium-high heat.

Dredge the fish steaks in flour on all sides and place them in the pan when the oil is hot, but not smoking. Cook until the fish is golden brown underneath, then turn the steaks and fry the other side until crispy. This should take about two minutes per side.

Serve straight from the pan with wedges of lemon, apples and pears. Have plenty of sweet lemony tea made (vodka shots if it is a special occasion) and be prepared for people to drop by once word gets out.

Makes a meal for six; more if you are serving it as a snack or starter.

Note: We strongly encourage using sustainable seafood for this recipe. The Madrid Protocol on Environmental Protection, signed in 1991 and entered into force in 1998, prohibits disrupting wildlife. While the kind of small-scale fishing a few of us did was not yet a breach in 1996, we are aware it was a grey zone and in hindsight are uncomfortable.

ROASTED BEET SALAD (RUSSIAN VINAIGRETTE)

I was so pleased with myself when I made a "Russian Salad" from one of the cookbooks I brought even after Lena enlightened me that she'd never had anything remotely like it in Russia. It turns out the combination of beets, cucumbers and hard-cooked eggs, tossed with sour cream is more Scandinavian in origin. A classic Russian Vinaigrette, which is what Lena called "their" beet salad, starts with beets and varies its extra bits from region to region and household to household.

Frankly, I don't know why you'd want to add anything to it—tossed with a simple French vinaigrette (that is to say, the dressing), the earthy goodness of the beet shines through.

2 pounds small to medium beets // 1 clove of garlic //
½ cup rice vinegar // 2 tablespoons honey // salt and pepper to taste //
2 tablespoons fresh dill fronds (optional)

Preheat the oven to 400°F. Wash and trim the tops from the beets—I like the way the ends taper so I leave them intact. Plus this minimizes "bleeding" while they cook. Wrap the beets in foil in groups of four and place on a baking sheet. Roast until they are tender and a paring knife slips easily into their centres, 20–25 minutes for small beets or 30–40 minutes for larger beets. Unwrap to cool.

While the beets are roasting mince the garlic and put it in a medium bowl with the honey and rice vinegar. Whisk and set aside.

When the beets are cool enough to handle gently rub off the skins. Cut lengthwise into wedges to make the most of their contours and add to the bowl containing the dressing. Mix well and let stand for at least an hour before serving. Season with salt and pepper and dill fronds if using.

Makes enough for six people.

KING GEORGE ISLAND SALAD (VINAIGRETTE)

Branding is everything. I hadn't planned on using tinned asparagus to make salad when I ordered it, but when I became disheartened by the condition lettuce arrived in after a long sea voyage, I gave up on salad greens entirely and reached for my can opener.

1 jar white asparagus spears (12.5 ounces) //
5 slices of bacon (more if you feel inclined—I was rationing mine) //
1 teaspoon cumin seed // 1 avocado // balsamic vinegar

Gently drain the water from the asparagus and carefully remove the spears so as not to break them. Set aside. Fry the bacon until very crisp and crumble it into small pieces in a paper towel. Put the cumin seeds in a small dry pan over low heat and toast them, shaking the pan gently from time to time. When you begin to smell their aroma and hear them crackle, transfer them to a small bowl. Halve and pit the avocado; slice into thin wedges then use your fingers to pull the rind away from the flesh. To assemble the salad, arrange asparagus spears and avocado slices on each of the plates. Sprinkle with crumbled bacon and cumin seed. Finish with a drizzle of balsamic vinegar or let your guests add their own.

Slices of hard-cooked eggs with chopped scallions or minced red onions make a nice replacement when stores of avocado run low.

Makes enough for three or four side salads.

A BIT ABOUT BEANS

Cannellini, navy beans, pea beans or any white bean will work in these recipes. For my pantry, Antarctic or otherwise, I default to great northern beans mostly because I love their name.

Soaking beans overnight is not strictly necessary, but you'll find it cuts down their cooking time and if you need them to remain whole for the recipe it helps prevent splitting. It's worth the fuss even for beans you end up mashing as it cuts back on gastric distress, something to consider whether living in close quarters or not.

Pick through the dried beans and discard any stones, fibres or funny-looking beans. Rinse under cold water, toss them in a bowl and cover with water. While ratios of water to beans are likely indicated on the back of the package, you can't go wrong with 6–8 cups water to 1 cup dried beans for an 8–12 hour soak. Rinse before cooking.

WHITE BEAN AND ROASTED GARLIC PÂTÉ

The tree-planting season I wrestled with a surplus of raisins and added them to unsuspecting recipes still haunts me (and probably some tree-planters) Not so the noble white bean. Dried or tinned, in soups or in stews, you can take it on expedition or serve it at a cocktail party. I love this pâté as an appetizer with crostini or to round out a good ploughman's lunch of cheese, dried sausage, wedges of fruit and warm bread.

1 cup dried great northern or some other white bean // 1 bay leaf // 2 heads of garlic //
6 tablespoons olive oil // 2 teaspoons dried rosemary // 1 teaspoon coarse salt //
zest of 1 lemon // cracked black pepper

Have the beans soaked and rinsed. (See page 123.) Put them in a pot with the bay leaf and cover with cold water. Bring to a boil, then reduce the heat and simmer, covered, until the beans are soft, about an hour or so—longer if the beans haven't been pre-soaked. Check on them from time to time to skim off any accumulation of foam. Drain and set aside.

While the beans are cooking, wrap the heads of garlic in foil and roast in a 400°F oven until the sides yield nicely to your touch, 15–20 minutes. Set aside to cool.

Heat the oil in a skillet over low heat and crumple in the rosemary and salt. (When I get the urge I mince a few additional cloves of garlic and add them at this stage.) When you begin to smell the rosemary stir in the beans and cook just until warm, about 3 minutes.

Set aside about a third of the beans. Squeeze the garlic into the pan and mash with the back of a fork to make a thick coarse paste. When the garlic and beans are thoroughly amalgamated and you are satisfied with the texture, stir in the lemon zest, cracked pepper and reserved beans. (You could also mash all the beans if you want a more refined texture.) Warm thoroughly and serve with crostini or warm bread. Garnish with fresh herbs if you have some.

Makes about 2 cups or a good-sized ramekin.

WHITE BEAN AND ROASTED GARLIC SOUP

I thought I'd never see the end of our dried white beans until it occurred to me use them to make more than minestrone soup and pâté. Paired with roasted garlic and puréed to creamy perfection, they make a soup I could never tire of. If you are feeling fancy, grate in a heel of Parmesan or Romano cheese, add chopped dried sausage or crumble in crispy bacon or prosciutto.

3 cups of dried great northern or some other white bean // 4 bay leaves // 2 onions //
3 celery stalks // 2 carrots // 1 large potato // ¼ cup olive oil //
1 teaspoon crumpled dried sage // 2 teaspoons dried rosemary // 1 tablespoon salt //
9 cups of water // 5 heads of garlic // whipping cream (optional)

Have the beans soaked and rinsed. (See page 123.) Put them in a large pot with one of the bay leaves and cover with cold water. Bring to a boil, then reduce the heat and simmer, covered, until the beans are soft, about an hour or so—longer if the beans haven't been pre-soaked. Check on them from time to time to skim off any accumulation of foam. Drain and set aside, reserving a cup of the cooking water.

While the beans are cooking, mince the onions and peel and dice the remaining vegetables. Wrap the heads of garlic in foil and roast in a 400°F oven until the sides yield nicely to your touch, 15–20 minutes. Set aside to cool.

Place a large Dutch oven or heavy-bottomed pot over medium-high heat and pour in the olive oil. When the oil is hot toss in the onions, celery, remaining bay leaves, herbs and salt. Reduce the heat to medium-low and cook until soft. Mix in the carrots and potatoes and sauté for about 5 minutes. Stir in the drained beans, the reserved cooking liquid and 9 cups of water. Squeeze in the roasted garlic. Cover and simmer slowly until the vegetables are tender, about another 15 minutes, then remove the bay leaves.

The soup is rustically lovely as is, but the flavours really come together after it is puréed. Sometimes I'm happy with the chunky version. When I want a combination of textures I purée two-thirds of the soup and stir in the reserved bit afterwards. Blending the whole lot produces such a sublimely creamy soup you'll likely find you won't need to add the whipping cream. If you do want to add it, now is the time. Stir in a cup or two just before you are about to serve.

Makes enough for ten to twelve people.

Penguin cookie cutter at Chilean Base, 1996

Russian team potato peeling, 1996

TRESPASSING

Carol

———

Back in Toronto I was negotiating where we would dispose of garbage collected at Bellings-hausen. Some to Argentina? Some to Russia? Maybe there is no morally superior place for garbage.

I was delighted to hear from happily returned Project Antarctica II volunteers. I also had to respond to the few who gave Wendy and team trouble. Some feedback was reasonable: they didn't get what they expected; they wanted to spend more time with the Russian biologists; cleanup equipment was lacking; they needed more rakes, gloves etc. However, the majority of volunteers self-selected well and knew what they were getting into which was that they didn't know what that was. To be honest I had no idea exactly what we would be doing either but only that we were part of some kind of greater movement. All people who came on our project were willing to work but a few still thought nature was there for them. I had a volunteer from New York in the pilot cleanup at the Polish station the year before who wrote on her feedback form, "Not enough penguins."

Wendy and I loved reading Bob's *Condé Nast Traveler* article "Trespassing on Eternity." (1996):

> It was Christmas Day in Antarctica, and I am passing out gifts I made from items I found during walks amid a landscape that has been described as the earth's last pristine wilderness. For Wendy, our cook, I have created a mobile consisting of a padlock, a bent fork, a tin can lid, the base of a lightbulb, half a hacksaw blade and the head of a small ax...There are other gifts too: a penguin made of corroded welding rod, necklaces strung with an assortment of rusted nuts and washers...Odd gifts perhaps. But we are an odd group: seven people who've paid a considerable amount of money to spend Christmas in Antarctica taking out somebody else's trash. It's not a bad investment, really. It allows us, for five days, to become one of the very few tourists who have ever lived ashore in Antarctica, as opposed to day-tripping from tour ships. More importantly it allows us to convince ourselves that we have an answer to the question visitors to Antarctica find themselves asking: "Do we really belong here?"

Volodya Cook on steps of the mess, Bellingshausen, 1996

Vladimir (a.k.a. Volodya Driver), Bellingshausen, 1996

THE DRAKE SIDE

Wendy

———

Doc Roberto seemed more excited about sharing yerba mate than the scenery the day we flew to the Drake. I wondered if there was something other than "traditional Uruguayan herbal tea" brewed in the cup he passed around the helicopter cabin. I'd read in a travel book to be wary of South American men offering drinks. This one—served in a gourd, with a metal straw like a mouthpiece on a hookah—made me uneasy enough to embarrass myself by asking if it was the hallucinogenic tea referred to in *Let's Go South America*. Thankfully the passengers didn't understand my question; Lena and I could tell they wanted us to like the tea from their homeland as much as the place they were taking us to.

The tea is hot, so it's good. From the air, we got a sense of how temperate the Fildes Peninsula is; the moss was bright emerald green in the sunshine. Small lakes—melt ponds we didn't know existed dotted the tip of the island we considered our backyard. North of Bellingshausen, in the highlands, snow remained in the valleys and on all north-facing slopes on the moraines running parallel to the glacier. Signs of summer ended abruptly at the icefield. Our pilot followed that line on the traverse to the Drake. Jagged blue ice and deep crevasses look far more menacing up close. It would have been foolhardy if Sean and I had actually tried to ski there.

Everything about the Drake side of the island is wilder. Icebergs and growlers crowd the shores and fill the waters, then nothing.

The beach was packed with elephant seal sunbathers, a family of fur seals too. I always loved the story Sean told me about how the first mariners to see fur seals thought they were black bears that had returned to the sea, because of their bearlike heads, grouchy disposition and the way they walk on all fours. The story came to life when a cow, teeth bared with flippers slapping on the ground moving more like legs, chased Roberto down the beach at full tilt and very nearly caught him.

Even the rocks are different. Some larger formations rise at close to 90-degree pitches and level off abruptly as if they'd had their tops sliced off. Startling bands of green and a deep earthy red push out from black bedrock—no one seemed to mind not knowing the names of the minerals that day.

"We 'three sporty girls'…beg of you to take us with you on your expedition to the South Pole. We are…willing to undergo any hardships that you yourselves undergo. If our feminine garb is inconvenient, we should just love to don masculine attire… We do not see why men should have all the glory…especially when there are women just as brave and capable."

—Peggy Pegrine, Valerie Davey and Betty Webster's letter to Shackleton, January 11, 1914. His reply: "There are no vacancies for the opposite sex on the expedition."

Jackie Ronne (first female working member of a U.S. expedition) and Sig Gutenko wrapping pemmican, Ronne Antarctic Research Expedition, Stonington Island, 1947

BROKEN ICE

Carol

I don't think Wendy, Lena or I cared much about the fact men outnumbered women in the Antarctic. Even before we boarded the icebreaker that would take us to Bellingshausen, we knew there would be a certain dynamic.

There is a long tradition of pioneering women to Antarctica—the first woman to work in the region was botanist Jeanne Baret, who visited the Falkland Islands in 1766–67, disguised as a man. A nightshade plant, Solanum baretiae, was recently named in her honour. In 1935 Caroline Mikkelsen, a Norwegian whaler's wife, was the first woman recorded to have stepped on the continent.

Admiral George Dufek, the first commanding officer of the U.S. Operation Deep Freeze of 1955–1956, said women would go to the U.S. Antarctic program "over my dead body."

The first female Russian in Antarctica was marine biologist, Maria Klenova, who went in 1956 and helped map the first Soviet Antarctic atlas.

In 1969 an all-women scientific team led by Lois Jones with Ohio State University's Institute of Polar Studies (now Byrd Polar Research Center) did geochemical sampling in ice-free valleys. New Zealander Pam Young was with them. So much for Admiral Dufek.

Women are respected scientists, artists, activists, explorers, support staff and more. Today they represent one-third of staff at Antarctic bases, lead and participate in game changing research, such as Susan Solomon and team who helped identify the cause of ozone hole. Chlorofluorocarbons (CFCs) were depleting the ozone layer protecting life from the sun's harmful ultraviolet light. Scientists and politicians acted following the discovery: The Montreal Protocol (1987) was a landmark international environmental treaty banning CFCs.

Barbara Hillary, a retired nurse, was the first Black American woman to reach both Poles (at age 75 the North Pole and age 79 the South Pole). She shared her mother's message, "make sure your own window is clean before you criticize others."

Wendy and I saw at Bellingshausen and while visiting other bases on this mini United Nations island in the middle of the ocean that no matter what the gender, class, rank, nationality or position of Antarctic scientists and visitors, everyone is interdependent.

NAVIGATION

JANUARY 16–31, 1996

Chocolate chips and brown sugar for cookies—finally!
And Russian lessons at radio house with Ilya afterwards.

JANUARY 22, 1996
WT   Deliver Hot Cross Buns

Lunch:
· White Bean and Roasted Garlic Soup, African Lentil Soup or Chowder
· leftover pizza
· fruitcake, muffins or cookies

Camp 8. Republican Senator Fred and his wife, Nancy. She is so lovely coming into the
kitchen after dinner to chat about my cooking influences. I like that neither of us could pin
down our national cuisines. What a poker face—she really had me going with that piece of
wood she said she found washed up on the beach—driftwood in the shape of a cowboy boot
with a Kansas sticker on it. Both fit and strong—with Fred's farming work ethic I think
they could have come close to wrapping the cleanup at Stoney today, but Sean's really trying
to break up the work there now. Smart guy.

   Sunny again, quite unusual two days in a row. Hiked over to Stoney to deliver lunch—
well at least part of the way. Seems impossible to walk anywhere. People spotting me on the
horizon, stopping to pick me up. Sergey and Vadim were hilarious, pulling up in the Niva
like they were just out for a little tootle around the island. Thwarted plans to lose myself in
a language lesson on Sean's Walkman. Scrambled up the big headland at Stoney on the way
back at least. Beautiful, peaceful, an escape. What am I here for?

Later: Football match with Uruguay on the flats in front of their station. Maxwell Bay and Collins Glacier sparkling in the sun make a spectacular backdrop. Our team comprises a Scot, a Canadian, a German and three Russians. Volunteers cheering from the sidelines: "Davaj, davaj, davaj!" ("Come on, come on, come on!"). We lost with me in goal. Great fun anyway—everything from the ride in the back of the truck, all huddled together, guys fussing over Lena and me to make sure we're comfortable and warm.

One pair of earrings from Bio Vladimir
Ten more fish
Spatz now (sleep)

Camp 7 Debris Collection (Jan. 15–19)
Stoney Bay Area 10
Bellingshausen Area 3: 1 barrel mixed

JANUARY 23, 1996
Same old, same old. Grey, mild, changing wind, occasional shower. All hands on deck to work through the fruit and vegetables in the shed checking for rot and soft spots. Got all my preserving done—most satisfying. Freezer now stocked with eggplant, zucchini, green and red peppers, beans, tomatoes, apples and corn—grilled, stewed, blanched or roasted. Soups, sauces and pesto too.

Potluck meals with the rest of camp. Most of the guys made an effort to try everything, others more tentative. Notothenia chowder a hit and Dima was sweet to compliment me on my vinaigrette. Chocolate mousse was the hands-down winner. Think I blushed when Nancy said it was lovely and light, as if it could compare to dessert at a state dinner.

Bit of a wait for the *Petrov*. No bubble bath, but biscotti this time and a visit with Kevin. Letter from Bob laced with basil—brilliant! Fax from P. Now, I'm not sure about this one but it reads a bit like a Dear John letter. Three pages of "come here, go away" and then, "I'm still missing you something fierce." He's twisted everything again. I said he should use the time to sort things out. I didn't say I wasn't coming home. Not sure how I'll respond.

Got white potatoes and that's the most important thing. What else? Chocolate chips and brown sugar for cookies—finally! And Russian lessons at radio house with Ilya afterwards—Dima, Lena and Radio Sasha as teaching assistants. Lena always makes a big deal when I'm good at a sound or word. "Zha, zha, zha." I'm not sure how useful it is to be able to say "I'm hot." ("mne zharko") as well as I do since I rarely am, but I'm glad I mastered "U nas net vody. Pozhaluystra vody." ("We don't have any water. Bring water, please"). Vassiliy is in for a surprise when I call tomorrow. I've got so many languages going through my head everything is coming out wrong. Still I'm ridiculously proud to have the Cyrillic alphabet down and today at lunch Sergey and Vadim said, "Vendi, you are now speaking as well as a three-year-old."

Bio Vlad's handmade scrimshaw walrus-tusk earrings, a gift to Wendy

Penguin carcass, Deception Island

JANUARY 24, 1996

Day off—much needed. Lunch with Chilean base commander and officers. Not as animated as the Christmas lunch, but got Fernando's recipe for cazuela. Invited to cook dinner for the lot of them tomorrow night. Not exactly my idea of a holiday, but I might learn something and Lena's keen to go. Wouldn't hurt Zuniga to invite Sean to one of these things.

Volodya Cook's yummy pancakes for dinner tonight. They're almost like biscuits. I had three helpings and lots and lots of cream. What luck to be immersed in a culture that celebrates Pancake Day for a week—February is looking good. Lots of joking; morale is high.

Made earrings with stones Sasha Diesel gave me. Paper clips for hooks.

No luck with Inmarsat. Will have to write to P.

Camp 8 Debris Collection (Jan. 19–24)
Stoney Bay Area 11: ½ barrel mixed waste & piping
Bellingshausen Area 4: ½ barrel mixed

———

What to give when you don't have anything—
Feathers, fish, stones of all sizes and shapes
A bouquet of lichen, shiny bits of metal and pins for my hat
Fruit from who knows where—the Chilean kitchen, the ships?

Our guys had gone without fresh produce for months; why give a portion to me?
Even with my love of lists it's not like me to keep a running tally of gifts. Perhaps I kept track to understand and to hold on to whatever lies behind the desire to give? It was a good impulse.
Now when I explain these strange and lovely rhythms I liken them to the way we often find ourselves drawn to the only toddler in a room. Or the way a male penguin will search a beach to find the perfect stone to present to a female and when he finally finds it, waddles over to place the pebble at her feet for her nest.
I think of the day Lena and I spied Anatoliy and Vassiliy up in the hills like shepherds with walking sticks and a large cloth sack. Hours later they came to the door at Canada House and gave each of us amethysts the size of butternut squash.
Tiger's eye and jade from South Africa, jasper, quartz and amethyst from King George Island. If Lena and I had been roosting penguins we would have had the most beautiful nests in the rookery.

JANUARY 25, 1996

Made dinner tonight for base commander Zuniga, Doctor Rolando and Felipe the young heli pilot just in from Punta Arenas. A real home with running hot water, a cd player, a microwave and sterling candlesticks of all things—still a little culture shocked.

Lena gave me a pelmeni lesson. My tree-planter-sized attempts were a bit structurally unsound—perhaps because I used mashed potatoes in the dough. Felipe said they were more like empanadas.

Pelmeni = perogies = pasties = vareniki = empanadas = patties = dumplings = potstickers.

Bit of a musical language lesson as well tonight; Zuniga translating Spanish love songs in time with Julio Iglesias while Doc packed back the candy canes we brought for dessert. Made dinner plans for when the new volunteers arrive plus a footy match and a Canada House barbeque in two weeks. Offered a ride home. Makes me laugh—it's only a ten-minute walk up my mountain. Zuniga asked if I'm scared living alone. Should I be? And then out of nowhere he volunteered his bathroom. Assured me it is private and says I can use it whenever I want. Second offer of shower from a base commander in one week. Funny. I'm not sure I'll get used to this special treatment. Maybe they think I'm swimming out of necessity, not choice— who knows? But I don't want to offend our guys by accepting or travel even farther to get clean, and really nothing can beat the hot sauna.

I didn't know there was another sauna, a hotter sauna, until New Year's. The Diesel guys were like boys in a toy shop that day, pointing out its proximity to my swimming hole, wash basins stacked under a marble bench, a wringer washer, clotheslines strung by the generator fans and a watchman to guarantee my privacy.

Three rooms just for getting clean—two of them lined with gorgeously aged tongue-and-groove wood panelling and lighting soft like candlelight. I asked about a tin tub, the shape of a coffin propped against a wall. One of the guys apologized about its manky condition— it hadn't been used in years. And they were all a little sheepish about an intricate mural someone burnt on the wall one long winter dreaming of tropical islands and women. I fell in love with the place immediately, and would have switched to the old sauna even if they hadn't promised to ignore the roster and crank the heat over 200° F whenever I want.

Some days it's cooking, others it's cleaning that affords me time for saunas. After morning chores and setting dough to rise, I head to the freezer for supplies and stop at Diesel to put food in basins to thaw—soup for lunch out front closest to the generator fans. I grab a sauna while I wait.

No dallying. Drop hose in tub. Draw bath.
The sauna always hurts, at least until I break a sweat and I have to stay very still or my hair will singe my back when it brushes against it. I never use the birch branches the guys leave out. When can stand heat no longer, pull up straps, slip on boots and Dad's hunting jacket.
Walk—don't run past generators. Dash to stream.
One quick plunge.
Repeat and cover with towel.
Wave to watchman on way back through Diesel.
Long, long soak in tub to warm up—quiet except for the muffled hum of the generator and the only pressing thing to think of is whether the soot will come out from under my nails this time. Shower to rinse—an option of pure decadence.
Dress.
Out to generator. Flick head over to dry hair by fan—look away if watchman watching.
Time permitting, stop for tea. Soup thawed. Back to work.

One rock from Radio Sasha.

JANUARY 26, 1996

Strangely calm. Barnyard smell from the Drake. News that the *Multanovskiy* bottomed out on a landing—wonder if that group of volunteers will come at all.

More jewellery-making with Sasha Diesel. Today he showed me the original mess and kitchen in the derelict buildings behind Diesel. Tinkering and a little sewing.

Must get to other projects: sketches, photographs, castings with penguin bones.

JANUARY 27, 1996

And what of today? Predictable weather change for a changeover day. As the guys say, winter is coming. New group looks good—let's hope so. Fax from Carol brought us down a bit as did mine from P. I'm glad I don't have Carol's job. Best to try to keep it in perspective, I guess. She's just been fielding calls from those men in Camp 6 who thought they'd been shortchanged because Sean isn't a capital N naturalist. To my mind he's got so much cred from his field-work with Scott Polar on Cuverville Island last season. Good thing they never found out I'm a cook and not a chef. And their preoccupation with whether the work they did was going to have a lasting impact. Bellingshausen certainly isn't a consumer society and Sean says given the debris they collected accumulated over 28 years, rates of littering are clearly low. Yesterday he inspected areas cleaned earlier in the season and found no deliberate re-littering. There are wood fragments and wire in places but they're only due to the larger waste removal efforts the guys recently initated.

P: How can I be dumped so many times by one guy? It's strange; I know we don't belong together. His comment, "You could get a job in an office if you wanted," still stings. But part of me loved what he offered. Am I sad because it wasn't enough? So what is? What is the trade-off? Do I expect too much?

Dinner: White Bean and Roasted Garlic Pâté, Zucchini Toasts with Tapenade, French Onion Soup, All-In Pizza, Caesar Salad, Custard with Fruit Compote in Phyllo Nests.

JANUARY 28, 1996

Lunch: Spinach Soup and Salade Niçoise, Pineapple Upside-Down Cake.

Wind and snow—a real winter storm and it keeps on. 1:15 a.m. now—still blowing hard!

Lost in thought and somewhat melancholy today. Lena sweet and comforting, "He wanted a pocket girl and you are not a pocket girl." And later, in her lovely, direct Lena way, "Whatever he has to say it's not about us. Don't read it. Don't think about him."

I remember a talk I had with Bob on Christmas Eve. His words hang out there now, "It sounds like you've either found the right person or you haven't." Guess not.

I'm sorry to be wasting my time on this. Good day otherwise. I like this group and the guys were in top form today. Volodya Cook and Vassiliy spotted me on my way to the freezer this morning and ushered me into the mess. They really put it on, tripping over themselves as if they were in a Laurel and Hardy skit: taking my coat, pulling out my chair, helping me with my napkin, serving me lunch, waiting tableside for my critique. Crepes with meat filling, watermelon and cranberry goop (too thick this time) and a spoonful of borscht. All yummy,

Lena Nikolaeva (in blue coat) unloading supplies from the Zodiac
arriving at Bellingshausen station, Ardley Bay, 1995

Sasha at radio call, 1996

but I spoiled my appetite for the lunch I was making. Bio Vlad stopped by to lend me his Walkman and Russian lesson cassettes. Swung by Diesel on my way back and discovered my laundry had been folded—later I found out that Volodya Driver had had words with someone about my underwear hanging to dry. Comical.

Zuniga and Felipe for dinner tonight. Love how they get into the spirit of things. Commander orchestrated a light show along the runway during dinner. What a love–hate thing I have with that airstrip—always spoiling my view of Flat Top when the clouds finally lift. Still, we all can't help but crane our necks when planes land, as if we might know someone getting off. And tonight the dancing lights made Marcia's 60th birthday! She tittered when they gave her birthday kisses both in Chilean and Bellingshausen time. Men in uniforms.

JANUARY 29, 1996

Beautiful sky. Windy and cold. Glum—I really want to talk to P and Radio Sasha couldn't take his chances pirating a call again. Maybe tomorrow. All was not bad or lost. The volunteers are shipping out tomorrow so they took a break from work. I tagged along to tour the Chilean hospital and school. Given chocolates, offered a sack of flour. Later walked to Great Wall station. Like stepping into China: mahogany veneer drop-leaf tables with ornately carved legs; bottomless pots of green tea. Mr. Wong let me poke around the kitchen while everyone else shopped and got their passports stamped. Giant wok over gas flames, huge rounds of wood for cutting boards; I could have stayed there all day. I'll get my chance when we swap recipes. Picked up laundry at Diesel. Tied with a ribbon so I know who did it for me. Dinner okay in spite of having no idea how to cook lamb ribs. Cake good. Earthquake during dishes— a few things tumbled off the shelves.

JANUARY 31, 1996

Absolutely outstanding day today. Sun and NO wind. Had our first barbeque at lunch on the barbie the guys made for me. Enjoyed this group of volunteers very much—the larger, the better. Made potato chips with all our different types of potatoes again. Kathy, the engineer, travelling with her dad joined me in the kitchen—helped select seasonings to dust each batch: cumin seed, cracked pepper and rosemary, cayenne and coarse sugar. Everyone else judged our experiments. *Petrov* late. A shame we just get comfortable with a group and they are gone. New group looks good. Note from Kevin that he couldn't set sail without my bubble bath. But we are happy to have a tape deck now even if we have only five cassettes among us: Tom Waits, Ani DiFranco, Katherine Wheatley, The Waterboys, Everything But the Girl. The piped-in Radio Moscow option was wearing.

Camp 10 Debris Collection (Jan. 27 31)
Bellingshausen Areas 5 and 6: 1½ barrels mixed

STONEY BAY CHOWDER

"Wendya fish!" over the telephone was the standard invitation to go to Diesel for fish. This continued for a long time after the guys learned how much I liked Notothenia. When I wasn't able to join them, they brought the latest catch to my door. I served Fisherman's Fish at lunch, or as a starter, and when the catch was particularly bountiful I made chowder. I'll never know if the reason they stopped deliveries to Canada House was in response to my request for NO MORE FISH or because of the day I brought Stoney Bay Chowder to a potluck. Corn—talk about a cultural faux pas. Europeans don't know what they are missing.

½ pound slab bacon // 2 onions // 1 celery stalk // 6 medium potatoes //
2 tablespoons butter // 3 to 4 ears of corn (about 3 cups corn kernels) //
2 teaspoons dried thyme // 1 tablespoon salt // 1 bay leaf //
2 pounds cod fillets, (or some other white fish) // 6 cups water //
1½ cups whipping cream // freshly ground pepper

Have the onions finely chopped, the celery and potatoes diced and the corn kernels cut from the cob. Trim the rind and cut the bacon into a small dice. Toss it into a heavy-bottomed soup pot over medium heat and cook, stirring often, until lightly browned at the edges, about 2 minutes. Add the onions and celery to the pot and cook over low heat until soft, about 5 minutes.

Melt the butter and stir in the thyme and salt. When the herbs have warmed thoroughly, mix in the potatoes and cook until they are slightly softened. Stir in the corn and let it cook for a minute or two then bring up the heat, add the bay leaf and pour in the water. Bring to a boil, cover, reduce the heat and simmer for about 10 minutes.

Have the fish cut into good-sized chunks ready to add to the pot when the potatoes are tender. Gently cook the fish until it pulls apart easily, another 4–5 minutes. Stir in the cream and cook just long enough to bring everything up to heat. Add freshly ground pepper to taste.

Makes enough for ten to twelve people.

ALL-IN PIZZA

Pizza is a personal thing, so it's often best to let people make their own. When I recognized the ice-breaking potential for this hands-on meal, I started to serve it the first night of each camp. I put out a stack of partially baked pizza crusts with a variety of toppings and let the volunteers and dinner guests do the rest. Make-your-own-pizza night encourages creativity, shapes conversation (even when there is little) and is a fabulous way to turn around leftovers.

Pizza Bases
1 batch Honey Oatmeal Bread dough (page 81) made through the first rising //
cornmeal for the pan

When the dough has doubled in size, turn it out onto a lightly floured surface, punch it down and cut it into four equal pieces. Knead each piece a few turns, roll them into uniform balls and set aside to rest, covered, for about 5 minutes while you grease your baking sheets and preheat your oven to 350°F.

To make pizza crusts that are the same shape and size, roll out a ball of dough into 14 by 14 inch square about ¼-inch thick. Cut four rounds from the dough using a 7-inch pot lid or a bread and butter plate as a template. Continue with the remaining dough. Sprinkle the prepared baking sheets with cornmeal. Transfer the rounds to the baking sheets.

Bake until the bases begin to brown slightly around the edges, 8–10 minutes. Turn out onto racks immediately to cool and repeat with the rest of the dough as baking sheets become available.

If you prefer the look of a more free-form pizza, divide your dough into sixteen pieces and shape each of them into a ball. Proceed with a rolling pin or use your hands to press and pull one of the balls of dough into a pleasing shape. Continue until you have formed and baked all your pizza crusts.

If you are going to use your bases later that day, they can sit out. If not, stack them in an airtight container or wrap them in plastic and freeze until ready to use.

Makes bases for sixteen pizzas.

Toppings

The "All-In" part comes into play with the toppings. In Antarctica I grilled eggplant and zucchini, made tomato sauce with and without sausage, and roasted red peppers specifically for pizza night. This may seem like a lot of work for a meal billed as a make-your-own affair, but not if you have a stash of sauce or grilled vegetables built up in the freezer.

The prep for other toppings often only involves opening a can, slicing a few fresh vegetables or pulling leftovers from the fridge:

· Canned shrimp, anchovies, asparagus, olives, or pineapple chunks dress up any pizza— perhaps not together.

· Slice some red onion, chop a green pepper, and grate a selection of cheeses.

· Pull some pepperoni, bacon, dried sausage, or cooked ham from your refrigerator.

· Leftover roasted potatoes or chicken of any kind make lovely additions.

· Tapenade or pesto add some zing and caramelized onions are sublime. In fact you could just stop at caramelized onions.

Assembly

Other than a suggestion to make structurally sound pizzas I keep directions to a minimum. Sauce goes on the bottom; cheese is the glue and anything that would be good crispy (bacon, sausage, salami, potatoes) should go on top.

Preheat the oven to 350°F as the pizzas are being constructed. Bake on baking sheets in the centre of the oven until the crusts brown up and the toppings have melded together, about 10 minutes. Have something for people to nibble while they wait for their creations.

Makes enough for sixteen people.

CARAMELIZED ONIONS

As far as I'm concerned you don't need anything more than caramelized onions, sharp cheese and black olives on top of pizza crust, but then you have to call your pizza pissaladière (French onion pie). To make pissaladière start with a generous layer of caramelized onions spooned on the crust, top with shaved Parmigiano Reggiano, or Romano cheese and stud with pitted kalamata olives. Bake in the usual way. Cut into diamond-shaped pop-in-the-mouth-sized pieces, it makes an elegant hors d'oeuvre. Or you can eat it all yourself.

15 onions (about 5 pounds) // 6 tablespoons olive oil // ¼ cup butter
6 bay leaves // 2 tablespoons sugar // 1 teaspoon salt // dry white wine

Slice the onions thinly lengthwise and set aside. Combine the olive oil and butter in a large heavy-bottomed skillet or saucepan set over medium heat. When the butter begins to foam add half the sliced onions and 3 bay leaves, then sprinkle in 1 tablespoon of the sugar and ½ teaspoon of the salt overtop. Lay the remaining onions overtop along with the remaining bay leaves, sugar and salt.

Cover, reduce the heat to low and cook for 20 minutes. Remove the lid, stir the onions and increase the heat to medium. Cook for another 15 minutes, stirring occasionally. Finally, increase the heat to high and cook, stirring constantly, so that the colour of the caramelizing sugar is evenly distributed and onions do not scorch. This should take another 10 minutes.

Pour in a splash of wine to deglaze the pan; the deep golden colour will even out among all the onions at this point. Cook for 5–6 minutes then remove from the heat and set aside. If you aren't satisfied with the colour, return the pan to the stove, pull the onions to one side and sprinkle a little more sugar on the exposed bottom. As it caramelizes bring it together with the onions. Caramelized onions keep for a week to ten days in your refrigerator.

Makes about 4 cups.

Rosemary-Crusted Lamb Ribs,
Red Cabbage Confit,
and Baked Stuffed Onions

ROSEMARY-CRUSTED LAMB RIBS

I come from a family of lamb lovers so I ordered plenty from our supplier in Ushuaia. I didn't expect it to be delivered intact, hoofs to the sky. Thank goodness for Volodya Cook who helped with the quartering and trimming.

You won't find lamb ribs in your grocer's freezer, so ask your local butcher. They may be called lamb ribs or lamb breasts and are well worth the hunt. I braise them a short time, top with a garlicky crust and finish them under the broiler. They are delicious alongside baked stuffed onions, mashed potatoes and red cabbage confit with something green to round out the plate, or try them with ratatouille and grilled polenta.

3 pounds lamb ribs // 2 bay leaves // 8 cloves of garlic // 1 onion //
3 tablespoons grainy Dijon mustard // 2 tablespoons red wine vinegar //
2 tablespoons each of dried rosemary and dried thyme leaves //
ground black pepper and salt to taste

Trim the ribs of any excess fat, but don't be too rash—the fat carries a lot of the flavour. Place the ribs in a large pot filled with water, a few bay leaves and a splash of vinegar and bring to a boil over medium-high heat. Simmer for 10 minutes or so to render some of the fat, then remove the ribs from the liquid and cool on a wire rack over a baking sheet.

Mince the garlic and onion finely and put them into a small bowl. Stir in the mustard and vinegar. Rub the herbs together between your fingers as you add them to the bowl—don't be afraid to get rough with the rosemary if it is a little twiggy. Add the salt and pepper and mix well to make a coarse paste.

When the ribs are cool enough to handle cut them into sections of three ribs each. Rub the lamb all over with the herb paste. Place the ribs onto a broiling pan. A cooling rack set over a rimmed baking sheet also works well. Broil at least 4 inches from the heat source, turning once after about 10 minutes. Continue to cook until they are brown and crisp all over, another 10–15 minutes.

Makes enough for four as a main or eight as starter.

RED CABBAGE CONFIT

Given that Antarctica is kitchen pest-free, I didn't pay much attention to the cracks in our storage shed wall until I noticed that one tiny sliver of sunlight had been shining on a head of cabbage with enough intensity to cause it to sprout. Cabbage appears frequently on provisioning lists, but I imagine ours is the only expedition ever to have it as a table centrepiece. Later I prepared this simple recipe I'd grown up with—red cabbage braised with apples, currants and spices. It is one of those dishes that grounds a meal so beautifully you can hardly believe it came together with so little effort.

1 medium red cabbage // 7 apples // 4 small onions // $\frac{1}{4}$ cup currants // 4 bay leaves //
1 teaspoon cracked peppercorns // 5 tablespoons sugar // 1 tablespoon salt //
3 cups water // $1\frac{1}{3}$ cups white vinegar // 3 tablespoons butter

Halve, core and chop the red cabbage and put it in a large ovenproof dish or roasting pan with a tight-fitting lid. Peel and chop the apples and slice the onions. Toss them with the cabbage, along with the currants, bay leaves, salt and peppercorns. Pour in the water and vinegar and mix well, then top with pats of butter. Cover and cook in a 400°F oven until tender. Check from time to time to turn the cabbage and monitor the moisture level. You can always add a bit of water or apple juice to bring it around, or leave the cover off for a while if it is too soupy. It should be ready after about $1\frac{1}{2}$ hours. Braised red cabbage brings a vibrancy to just about anything. I love it with pork, beef or turkey—or with lamb, especially in the absence of my Mom's mint conserve.

Makes enough for six to ten, depending on the number of cabbage-lovers at the table.

BAKED STUFFED ONIONS

Cooking onions go the distance, yet so often they are consigned to soups and sauces, minced and chopped beyond recognition. I admit my motivation for weaving them into the menu as a side dish early in the summer was to make the other vegetables in my storage shed last longer so I wouldn't have onions and only onions to work with in our last few weeks. Stuffed and baked they make a refined addition to any meal even when you aren't dealing with a surplus of onions...or dried apricots.

6 small onions // 6 dried apricots // 1 apple // ½ teaspoon dried crumbled sage //
1 teaspoon dried rosemary // ½ teaspoon salt // 2 tablespoons butter //
2 tablespoons bread crumbs // ground pepper to taste

Place the whole unpeeled onions in small saucepan and pour in enough water to cover them. Bring to a boil, then reduce the heat and simmer until the onions are tender, about 10 minutes. Rinse under cold water until they are cool enough to handle. Trim the ends slightly and cut the onions in half crosswise, then carefully peel off the skin. Starting with the centre, gently pull away the layers one at a time until you are left with a shell about ¼-inch thick.

Mince the bits you've removed and put them in a mixing bowl. Finely chop the apricots and the apple and add them to the bowl with the herbs and breadcrumbs. (I like to keep the apple peel on for colour.) Melt the butter and drizzle it overtop. Mix well. Fill each onion shell, mounding it slightly, and place in a lightly oiled baking dish. Bake in a 375°F oven until golden, about 30 minutes.

Makes enough for six people.

CUSTARD WITH FRUIT COMPOTE

I was always frightened of making custard until our provisioning challenges forced my hand and it was the only thing I could prepare with what was available. Admittedly, my inaugural attempt in Camp 1 was a little eggy, but I eventually achieved a perfect creamy balance of ingredients sometime in January. Inspiration for the compote came from a 20-kilogram bag of dried fruit Sergey gave us and the canned fruit I ordered.

Custard with Fruit Compote is a stylish pairing, made with such modest ingredients. It may make you rethink how you stock your household—at least in the winter.

For the custard
1½ cups table cream // 1½ cups whipping cream // ¾ cup sugar //
6 egg yolks // 1½ teaspoons vanilla

Preheat the oven to 325°F. Have six eight-ounce ramekins or custard cups set in a roasting pan. Combine the creams in a heavy-bottomed saucepan and place the pan over medium heat; cook just until bubbles form around the edges.

While the cream is coming up to temperature whisk the egg yolks with the sugar in a large bowl until they are well blended. Whisking constantly, gradually pour a thin stream of hot cream into the eggs. Finish with the vanilla.

Ladle the custard into the ramekins. Skim off any foam. Add enough boiling water to the roasting pan to reach halfway up the sides of the ramekins and carefully set the pan on the middle rack of the oven. Bake until the edges are set but the centres still jiggle, 30–40 minutes. Transfer to a rack to cool. If you are making the custard ahead, cover and refrigerate for up to two days. Bring the custard up to room temperature before serving.

Makes enough for six people.

For the fruit compote
½ pound dried apricots (1⅓ cups) // ¼ pound dried pears (1 cup) //
¼ pound dried apples (2 cups) // ½ pound dried pitted prunes (1⅓ cups) // ½ cup raisins //
2 cinnamon sticks // 6 to 8 peppercorns // 2 bay leaves // 6 allspice berries //
3 cardamom pods // 4 cups water // ¼ cup honey // a can of sliced peaches (28 ounces)

Make the compote while the custard is cooking or up to four days ahead.

Put all the dried fruit in a large stainless or enamelware pot. Break up the cinnamon sticks; crack the cardamom pods and add them to the pot along with the peppercorns, bay leaves, and allspice berries. Pour in the water and drizzle in the honey. Bring to a gentle simmer over medium-low heat and cook until the fruit is tender but still chewy, 5–10 minutes.

When the fruit is almost fully cooked drain and rinse the peaches and slip them into the pot for the last 2 minutes. Allow the compote to cool completely. Transfer to a bowl or large glass jar, cover and store in the refrigerator for up to a week. The longer, the better. Serve at room temperature spooned onto the custard. Fruit compote is also delicious spread on toast, stirred into hot cereal or as a topping on pancakes or waffles.

Makes 2 quarts.

"The primus is started and over it is placed the cooker with annulus and boiler filled with snow...The other two have settled the dogs and arrive in time to receive a steaming mug of hoosh, which is nonetheless welcome on account of such adventitious ingredients as reindeer-hair. This ubiquitous constituent of all food preparation or sledging journeys is transferred in some mysterious way of molting sleeping bags."

—Douglas Mawson, *The Home of the Blizzard: Antarctic Expedition, 1911–1914*

Shackleton and Hurley skinning a penguin, Patience Camp, Frank Hurley, 1915

I LOVE PHYTOPLANKTON
Carol

———

In 1821 Russian cartographer and naval officer Fabien von Bellingshausen, the first person reported to have sighted the hypothetical *Terra Australis*, was dismal about Antarctica's future with its dark nights and cold climate.

Bellingshausen didn't foresee a continent of model international cooperation as realized at a research station named after him where volunteers would one day clear garbage left by the Soviets and a Canadian cook would swap a Brazilian recipe with a Russian glaciologist.

Explorer Apsley Cherry-Garrard popularized Antarctic literature in his 1922 memoir, *The Worst Journey in the World*, nine years after his return. On the Cape Crozier journey of Scott's *Terra Nova* Expedition (1910–1912), Cherry-Garrard carried three unhatched Emperor penguin eggs across the ice during sunless winter days and –60°c weather.

Cherry-Garrard carted those eggs home to the U.K., hoping to prove an evolutionary link between dinosaurs and birds. A dinosauresque phase of development in the embryos wasn't discovered. However, penguin skins collected a hundred years ago proved vital to scientists in the 1960s. They tracked the role of DDT to demonstrate the impact of pollutants travelling by ocean currents, even to pristine Antarctica.

Explorers Mawson and Shackleton also reported that penguin eggs, with their transparent jelly-like yolks, made a good omelette. Somewhere I read a historic tip to add a shot of whiskey to the penguin egg omelette.

Nowadays Antarctic visitors are prohibited from eating local wildlife. In any case, we might worry about any toxins in them caused by pollution.

Phytoplankton, microscopic creatures, are the foundation of our aquatic food chain. Our oceans are warming and warmer oceans produce less phytoplankton. At both Poles and in between, the health of the food chain is our survival.

I often thought about my next meal on King George Island. We were in good hands.

WOMEN'S SUMMER

FEBRUARY 1–15, 1996

Walk to Ardley Island with Dima, Lena, Hilltop Sasha and Sean.
Wait for tide to go down. Finally low enough to cross.
New hole in boot, sock off left foot, water so cold it makes me dance.

FEBRUARY 1, 1996

WT   Found an orange in my sock this morning.

Another beautiful sunny day today. Cloud in the afternoon, then more sun. Makes me feel badly for everyone who had to tough it out working in the fog and sleet last month. It will be even better when the rake order comes in so folk won't feel like hedgehogs rooting between the rocks anymore. Extra tired tonight so I'll be brief.

This group appears to be an easy one. Easy conversation. Easy planning. Three-hour excursion to Drake. Saw lots of fur seals, one crabeater and of course plenty of elephants. Things have dried up considerably over there, no more boot-sucking mud. Finally ditched my leaky wellies for my hiking boots—like ballet slippers. Laundry and shower today in free time—actually had some. Visit at Diesel—first in a while. I'm not really conscious of missing visits. Guess they are. Sasha asked me how the meat was, how well the freezer is working, as if he thinks the only reason I come for tea is to thaw meat.

Talked with Tolya and Sasha (Sergey for a while) about all manner of things. Childlike, it's our usual dialogue. Using maps, pictures, diagrams and whatever props we can find, we talk about where we live, who with, what it's like and how we'll have to visit one another. And today we covered my boyfriend status, my hair and how it's like springs—learned the word for beautiful ("krasivyj").

Sergey made a huge fuss, like an uncle looking out for a favourite niece, scolding them for neglecting to teach me sooner, as if to say beautiful should have been right up there with please and thank you.

Offered job cooking on ship back to Russia.

Invited for fish tomorrow as Tolya's going jigging.

Successful dinner. Sad Vlad and Hilltop Sasha came. Sad Vlad sang Russian folk music.
Sean will have to think of another nickname for him if he ever branches out from ballads.

Nice to have music around.

Get recipe for Sad Vlad's mother's pickled peppers.

FEBRUARY 2, 1996

Sun in the morning over breakfast, then socked in as doing dishes. In mist and cloud for rest
of day. Successful day. We like this group very much. It's so hard to know what makes one
group good and another bad. An excitement maybe? A sense of privilege? I don't really know.
Had four dinner guests tonight from the Chinese station. They came bearing gifts for each
of us—handbags, crackers and Tsingtao beer. So kind, but the real gift for me will be learning
about Chinese cuisine during our next break. Hope we can also arrange to visit during
Chinese New Year's.

———

*Garbage-picker or land steward?—I cringe when I read that "good or bad group" line in my journal
even though I know it stemmed from a changeover day fear that someone in the incoming boatload
would be as difficult to please as the two men in Camp 6. You might think we sat around assigning
pass or fail grades to each person.*

*There's always something. I liked it when volunteers weren't put out by dish duty. I've always run
an equal-opportunity kitchen and cleaning up after ourselves was the reason we were there in the
first place. If only I'd thought to award a gold star to the two women from Washington who initiated
the drip-dry dish sculpture contest we used to motivate subsequent camps.*

*Lena was always more at ease when workers were robust. She worked alongside, plucking bits
of debris from between friable rocks, even in driving sleet. But seeing others, especially any seniors,
stooped over all day weighed upon her.*

*With good walkers we could share the trek to the Drake side. Good talkers—top of the class.
We only had a few short days to make each group coalesce.*

*Talking about food would draw people out. Two soft-spoken Hawaiian women Sean had charmed
and teased about wearing work socks with flip-flops finally opened up to me with a recipe for beef
jerky. Hawaiians serve it with beer. Peter, an IT man from Texas in his late 30s, made and sent me
labels for my kitchen appliances that arrived before I left Bellingshausen. "To the Ends of the
Earth" is still affixed to my Cuisinart. I bonded over baking with Carolyn, a retired publisher from
Seattle. True to her word, an envelope full of mouthwatering recipes was waiting in my mailbox
when I arrived home. I kick myself for not pressing the South African couple for a recipe.*

*I wish everyone had been able to withstand the hike to the Drake side. Conversation flowed on
the way up the road past Marsh airstrip, but with a tenor of uncertainty that always made me smile.
Not many "nature walks" start at runways. By the time we reached the crest of the hill and started
overland, people were speechless. Flat Top towers over all the other rocks and as you get closer
the brown lumps on the beach turn out to be elephant seals in their wallows. If people were able to*

Hand-collected nails from Bellingshausen station in 1996

go farther we'd walk the shoreline, past the wallows, past where the tide drops brash ice sculptures, to the base of Flat Top. Part of me always wanted to scale it, lie down on the rocks where it levels off, feel tiny and enormous at the same time—mostly I loved that it was utterly inaccessible and would stay that way. I felt bad interrupting our silence to press on, but up and over some rocks was a whale graveyard and if you've come that far it's worth it. Solemn and life-affirming, the beach was scattered with stark white bones and a carcass Doc Sasha said had been there for two years, yet there was very little, if any, decay and no odour at all—not counting the elephant seals, of course. I could have spent hours there if it weren't for dinner.

We could tell if the project was beginning to have more meaning by discussions on the way back to camp. To hear words like fragility and responsibility, opportunity and awareness was always worth the wait.

Over the summer a handful of volunteers questioned our focus and whether the work would have a lasting impact. But these doubts were mostly expressed on trip evaluations, filled out just before boarding the ship. I wish those conversations had come up at the table in between swapping recipes. I'd know to ask, what about scientific research? What about art? No one knows what the outcomes will be; still we persevere. You wouldn't stop cleaning your kitchen because it might get messed up again.

Three months after I returned home, Herb from our final camp wrote before venturing to China, that he'd "never been to a more perfect, more basic place." Two years later Brigid, a 70-year-old with beautiful British skin—the type of warm and adventurous grandmother I'd like to be—wrote it was "a time to be remembered forever that has created a deep well of peace to be visited at will."

· *Carolyn from Seattle: Apple Crisp, Ritz-Carlton Scones, Zucchini Carrot Casserole, Plum Pie Baked in a Brown Paper Bag*
· *Fred from Alaska: BBQ Pork, Thai Shrimp Curry, Grand Marnier Turkey, Whiskey 'Shrooms', Ginger Tea, Margarita Chicken, Honey Basil Carrots, Champagne Ginger Shrimp*
· *Naomi and Ami from Hawaii: Beef Jerky*
· *Carolyn from L.A: Amaretto Mousse*
· *Bernice from Illinois: Decadent dessert called More*
· *Arnold and Linda from L.A.: Chinese Pepper Steak*
· *Elaine from Michigan: Lentil Apricot Soup*
· *Kathy from Pennsylvania: Chicken Couscous*

FEBRUARY 3, 1996
Cookies to Chinese.

Interior of China's Great Wall Station (Changchen), King George Island, 2.5 km from Bellingshausen, 1996

FEBRUARY 4, 1996

2:50 a.m.; still night. First light comes a little later. Horizon only pink—forever stretches forever here.

Got to get this down before days run into one another again. Usual kind of day yesterday—wind, sun, cloud, fog, sun—a King George smattering of everything. Days seem to be getting better. Feeling relaxed and able to do more. Spent morning writing.

Sauna-day chat with Fernando by my swimming hole after my first plunge. The shock of the water was so great my knees buckled when I got out so he challenged me to go in again. Caught up with him later over tea with Sasha Diesel and Tolya. Usual discussion of places we've worked, what I will do afterwards. Fernando wants to come to Canada. Discussion of capitalism. Are we capitalists? Sasha: no, because his hands are dirty. Fernando's defence for clean hands is that he is a cook. Their faces brightened when they looked at mine as if they were relieved to see grime embedded in my cuticles and knuckles like dirt clinging to a carrot. Love conversations with these guys. Always with illustrations, apologies for not speaking well, gesture and laughter. Light-hearted and meaningless. Meaningful because something has been exchanged, some common ground reached. Like when Volodya Driver toasted the Canadian Arctic and the Russian Arctic at New Year's.

Peculiar chat with him earlier. I think he told me he needed a prostitute—kind of spooked me. Talk after sauna weird and full of innuendo too. "Vendi you have lots of friends up to Canada House." Must make sure of balance.

Celebrated Sad Vlad's birthday in the mess hall. Everyone showered and dressed in clean clothes for the occasion. Special meal. Wine and vodka stores opened for the celebration. I took braided bread, asparagus pâté and chocolate cake. Drinking led to more attention than usual. Volodya Bio, hair greased back, looking like John Travolta on the make, blowing me kisses from across the dining room during dinner. Volodya Cook stealing one as he served me and then announcing to the room that I kissed him. Everyone at the elder's table roared "Vy?" ("You?") The usual suspect persisting. Sorted out by Volodya Driver and his left hook. First violence seen on base, but it's not dwelt upon. Seems he had it coming. Radio Sasha said, "Everyone felt it was so." Still, it makes me sad—sad and exhausted. Lena says Sergey is happy we "know how to behave, that we are good girls"—wish he'd share that with some of these guys so we don't have to put up boundaries.

Maybe it was my fault for not taking a stand, but I think something else must have gone down. Some boys-will-be-boys thing more difficult to comprehend than Russian. I mean blowing kisses is pretty harmless—hilarious in fact, given his addled state—and I thought I had it under control by ignoring him. I still think I did. No matter, it was over in less than a minute. Just as V tried to close in on me, the other V swooped in and shoved him across the threshold into the vestibule. Harsh voices, coats flying, the slam of a body against the wall —I swear the room shook—and then skin smacking skin, like nothing I've ever heard before. I realize I've been a bit sheltered. Hilltop and Radio Sashas steered me out of the mess so quickly I didn't get to see Bio Vlad's eye, but Volodya Driver's hand looks pretty gnarly and swollen. The Canadian thing to do would be to apologize, I guess—but what for—being a woman?

Seal skull, the Drake Side, 1996

Elephant seal wallows near Flat Top, 1996

Party moved to Sad Vlad and Ilya's place. Guitar playing and singing. Sad Vlad—that beautiful melancholy Russian ballad Carol sang. Radio Sasha—upbeat and fun.

Guest list expanding: Vlad, Ilya, Dima, Hilltop Sasha, Sean, Doc Sasha. Later Tolya and a Catholic priest from Chile who Lena says, "Is not very much a saint." Merriment. Very much like tree-planting, Lena and I dancing every dance. We agree it feels good to be held but maybe not as tightly as Pepe the priest does. Surprised by our guys' schoolboy timidity and trembling hands. Good thing for jiving—keeps the idea of closeness alive and no one loses an eye. Sean crashed early and missed out on all the fun—strange how he keeps to the outer limits.

Walked home at 4:00 a.m. Bright and quiet. Stopped by Diesel to see if they were still awake. Lousy sleep.

All quiet today after the party. Sat on the hill that hides Canada House from rest of camp listening to the waves on the lake and terns playing in the hills. Dinner downstairs again and afterwards a vintage Russian film I think I snored through. Dima pulled me out to see a spectacular sunset—if only we had a Zodiac for a cruise. Finally to sleep.

Camp 11 Debris Collection (Jan. 31–Feb. 4)
Bellingshausen Area b6: 1 barrel mixed

FEBRUARY 5, 1996
Berg in Maxwell that has been with us for at least a week, moving back and forth across the bay, is finally breaking up. Brash ice surrounding it and more spreading out across the bay. Covered in penguins now.

FEBRUARY 6, 1996
Beautiful sun, then cloud, then sun—got a tan today.

Walk to Ardley Island with Dima, Lena, Hilltop Sasha and Sean. Wait for tide to go down. Finally low enough to cross. New hole in boot, sock off left foot, water so cold it makes me dance. Visit Germans who are busy breaking camp, having finished their two months' research in the rookery. Ardley is more beautiful than I imagined. Bluffs and rocky spires like an installation of Giacometti sculptures, especially that big one that looks like a head, on the south side of the island. Finally able to see our berg up close and the turmeric-coloured moss that looked like fields of wildflowers from across the bay. Lots of penguins. Chicks huge now, almost the same size as adults. Parents exhausted out getting food all day—chicks running after them, the chase teaching them everything they need to know for life after they fledge. Flotsam on beach. Some real treasures. Too big to carry. It killed me to leave that wooden ladder there. Charming enough not to be tossed in with the debris the volunteers find in the dumpsite, but not so old as to be confused with anything that fell off the *Terra Nova* and should be shipped to a museum

And what else of this day? Printer not working properly as another converter bites the dust. Seems I'm not meant to send this fax to P, get some sense of closure. Set up Zodiac with Sad Vlad and Sean. Why, I don't know since it seems unlikely we'll get an engine. Spent evening patching my jeans while watching a Russian flick. Walk up the hill—wind and moon and mist. First time I've really noticed the moon.

Vassiliy, Ilya, and Anatoliy, Bellingshausen, 1996

Beach in front of China's Great Wall base (King George island)
2.5 km from Bellingshausen, 1996

FEBRUARY 12, 1996

Our luck that the *Multanovskiy* went aground and had to be dry-docked in Ushuaia. Schedule thrown off. No volunteers from the night of Feb. 3 to Feb. 7. Bubble bath arrived on the *Petrov* after a six-week wait. Nice break. Wondered if this is indicative of February weather. Lots of sun. Warmth. Everything drying up.

Rain and cloud returned today. It will be hard to get used to this weather after our women's summer—I suppose we'd call it Indian summer. Lena explained that after the harvest, after all the work is done, women could relax in the calm fall days. Love Sergey's version: "Women's summer named for the hearty warmth of women." He says all the right things.

Trip to Artigas in the Niva. It makes me laugh how well Sergey gets around in this car when I hear stories about men at better-funded stations getting their swanky four-wheel-drives stuck, making distress calls to our guys to pull them out.

Chef Henry was a little flummoxed when we arrived, as if I'd asked him a trick question when I requested a Uruguayan recipe. The recipe the officers have all been talking about is for an asado. Seems just like a barbeque to me. Second in command, Quique, showed me their setup and explained that traditional Uruguayan asados are done in the countryside with an animal carcass splayed on a cross, but no one had a recipe. They looked defeated, as if they'd let down their country. I am more intrigued with Henry's stuffed chicken (pollo relleno) anyway. Will try it tomorrow. Doc Roberto was sweet to ask Lena and me when we want to fly again. Don't know why Quique calls me Brujita (little witch)—must be a term of endearment?

Lunch:
· Minestrone Soup
· Open-face chicken & cheese sandwiches on caraway onion rye
· Dinner
· Roast Beef Tenderloin au Poivre
· Pasta Alfredo with Peas
· King George Island Vinaigrette with Eggs
· Raspberry Fool with Digestives

FEBRUARY 13, 1996

Fog and mist, though I was unaffected as I got outside only to go to my food cache. Frustrated today and wondering if I'll get all done that I need to do. Perhaps I'll feel better after I get these letters and postcards in the mail. Really must write Gzowski, do some drawings, at least get the colours down. Should go for a walk along the ridge, around the lake. Get grounded. Do something.

FEBRUARY 14, 1996

Grey. Light snow. Flurries. No wind. Smell wafting in from the Drake heavy over camp, like some ecological disaster, but it's only the elephant seal wallows. Busy changeover day. Clean up Canada House and ourselves. This Valentine's Day received gifts from the volunteers, now no fear of running out of shampoo and conditioner or hand cream—a veritable spa. Two lovely retired teachers from Toronto; they aren't very nimble so the work is pretty hard for them, but they are troupers. It helps that Sean tailors the tasks to each group and Lena is attentive, making sure everyone is warm, dry and taking regular breaks. Am tired tonight, but happy that my postcards are on their way home. Lack of communication was getting me down. Invitation aboard *Livonia* to give a talk about the project lines up nicely with the two-month anniversary of our arrival. Ship seems civilized. Sean spoke well and I was treated to a tour of the kitchen. The crew said they would all love to be doing what we are doing.

Back at camp new group seems good: seven this time; young and environmentally aware with energy to match. Warned by the crew about Arnold the Wanderer and how he disappeared on an excursion and they looked for him for over an hour. The last thing we need is someone deciding to play with an elephant seal bull or head off to the glacier. We'll sic Sergey on him if we need to. Dinner yummy: Uruguayan Pollo Relleno and Chilean Cazuela, Barley Bread, Chocolate Fondue. Heard that my bread has created a buzz on the ships. Nice to know you don't have to plant trees all day to like my cooking.

Camp 13 Debris Collection (Feb. 10–14)
Stoney Bay Area 12: ½ barrel mixed
Bellingshausen Area 8a: 2 barrels mixed

> Sally's Galley is the affectionate nickname given to the old canteen at the Pole...
> I stepped up to the counter, picked up a plate, and ran my eyes over the mountains of eggs and bacon and sausages, pancakes with syrup, and the makings of those oversize burritos...Most of the people around me who were tucking into those huge platefuls of breakfast were about to start nine-hour shifts on the construction site, hanging iron, welding, and hammering in that bitter cold air outside, or tougher still, carving out tunnels under the ice...using those big loggers' chain saws. It was reckoned that each worker down here would put away about 5,000 calories' worth of groceries a day—and even so, most of them would lose 15 pounds or so over the course of the summer.
>
> —Roff Smith, *On the Ice: An Intimate Portrait of Life at McMurdo Station*, 2005

FEBRUARY 15, 1996

Bergs last a week in Maxwell.

Moss turns colour overnight.

And we develop new ways of measuring time and days. Great walk to the Drake this a.m. Picked up penguin parts; hope I get time to make something with them. This is an energetic group. Feeling that I can't really keep up with them, this constant extending of oneself without time to refuel. People friendly, wanting to help...and I want to withdraw. I don't want to interact with anyone, not even the Russians.

This arm's-length thing is being tested even more now. But I had to laugh when Lena told me the guys in radio hut have started to assign case numbers to those exhibiting bush fever symptoms. She confessed this latest "case" hits on her too, but it doesn't make it any easier to navigate. He says he just wants to have tea with me, but oh the way he hovers. And why did I have to be twisting strands of dough into braids when he stopped by yesterday?

"You have very skilled hands Vendi."

"Yeah, I really like making bread. Thanks for the visit. Gotta get to my soup now."

Poor guy; at least he apologized this time. "I'm very sorry. I just very much like see you smile."

Incommunicado—It's so easy today to take for granted our capacity for instant communications. On King George Island, our most viable option was to forward letters on supply ships, cross our fingers and hope they reached home before we did. The Chilean base commander floated the offer of a personal fax machine, but it seemed fraught with complications. I steered clear. A moot point all the same as I rarely managed to draft anything more than a scribbled postcard or skeleton of a letter. Still, the impulse to write served to anchor me in place and simultaneously connect me to home.

As it stood, the time saved by faxing correspondence was negligible. Incoming faxes were ferried from Ushuaia by the next outgoing ship, arriving on King George after whatever circuitous tour of the Antarctic Peninsula the new batch of volunteers had taken. Though marked PERSONAL AND CONFIDENTIAL, they were neither. Carol divulged years later she'd heard my two 'Dear John' faxes from P were read by each subsequent handler long before I received them.

Of course there was an expensive and notoriously unreliable payphone at the Chilean base. I opted instead for Radio Sasha's secret wizardry with the phone at Bellingshausen. It was a radio phone, so you had to signal you were finished speaking like the walkie-talkie speak we used when communicating with the ships and one another: "Sean, Wendy. Sean? Do you read? Over. When are you breaking for lunch? Over" Naïve as I may have been, I sensed this option involved some sneakiness, so when the international radio operator requested our location during a New Year's call, I fudged the coordinates and replied that we were on a ship in the North Sea. Sasha's relief was palpable. I baked him a chocolate cake.

In the week bookended by P's faxes, Sasha twice tried to place calls before announcing, "No more are possible." Safeguarding himself from the operator's wrath or protecting me? My gut says it was the latter. It was the most loving thing someone could conspire to not do for me.

CHOCOLATE CHIP COOKIES

Soft brown sugar and chocolate chips are not easy to come by in Antarctica. I generally stick to the maxim "if we don't have it, we don't need it," but when it came to baking I couldn't compromise. I radioed one of the ships to see if they could part with sacks of the sweet stuff—it's funny what you can't live without. These may be the best chocolate chip cookies baked south of 60°.

1 pound butter at room temperature // 1 cup white sugar //
3 cups packed brown sugar // 1½ teaspoons each of vanilla, baking soda and salt //
3 eggs // 4 cups chocolate chips // about 6½ cups all-purpose flour

Put the butter into a large bowl and break it up with a wooden spoon or spatula before adding the two sugars. Cream until fluffy, then add the vanilla, baking soda and salt. Add the eggs one at a time, mixing vigorously after each addition. Stir in the chocolate chips followed by about half of the flour. Continue to add flour gradually until you can no longer stir it by hand. Dust your hands with flour and use them to lightly work in what remains—depending on the season and humidity you may not need it all. When I can hear and see a few chocolate chips liberating themselves from the dough I know I've added enough flour.

Preheat the oven to 350°F.

To make large cookies, break off as much dough as you would use to make a snowball and form it into a sphere. Gently press it into a flat disc the diameter of a hockey puck, about ½-inch thick throughout. A good trick to keep from overworking the dough and creating tough cookies is to imagine the dough is hot to touch.

For smaller cookies, divide each large sphere into four equal-sized pieces before you form the discs. If you are careful while shaping you should be able to make them beautifully symmetrical. Place the dough onto a greased baking sheet about 2 inches apart.

Bake in the middle of your oven until the edges are golden brown and the centres slightly under-baked. The large cookies will take 15–20 minutes; the smaller cookies 10–12 minutes. Allow them to set on the baking sheets for about 5 minutes, then transfer to racks to cool.

Makes 2 dozen large cookies or 8 dozen small cookies.

Prolong the lifespan of your cookies by freezing a quantity after they have been baked. If you don't feel like baking the entire yield, shape the cookies and layer the discs between waxed paper in an airtight container and freeze. To bake, place the frozen dough on a prepared pan and follow the usual cooking method.

ASPARAGUS PÂTÉ

White asparagus may seem pretty fancy for an expedition, but artichoke hearts were even more expensive, and I wasn't about to head south without a perennial crowd-pleaser like the creamy garlicky goodness that is Artichoke Pâté. I ordered a case of tinned asparagus in its stead. Asparagus pâté is every bit as decadent. No apologies.

6 cloves of garlic // 1 brick of cream cheese (8 ounce) //
1 cup mayonnaise // 1 cup grated Parmesan cheese // juice of half a lemon //
1 jar white asparagus spears (12.5 ounces)

Preheat the oven to 400°F.

Peel and mince 5 of the garlic cloves and put them into a food processor fitted with a chopping blade. Add the cream cheese and pulse until smooth. This will be faster if the cheese is at room temperature. Add the mayonnaise and pulse until it has blended with the cheese. Add the lemon juice and all but 2 or 3 tablespoons of the Parmesan cheese. Pulse until the texture is consistent throughout, and then transfer it to a medium size bowl.

Drain the asparagus and chop it into small pieces. Add it to the cheese mixture and mix until evenly distributed. Scrape the mixture into an ovenproof dish. Top the pâté with the reserved Parmesan. For a crowd I often bake the entire yield in one dish. I also like to divide it among three or four small ramekins so that I can freeze some.

Peel and slice the remaining garlic clove into thin wedges and press the slices into the top of the pâté. Place in the centre of the oven with a pan underneath to catch any drips. Bake until golden brown and puffy on top, 10–15 minutes for small ramekins and 20–25 minutes for a larger quantity. Don't be discouraged if the top starts to look like an environmental disaster—this recipe can be a little finicky. You can salvage the pâté just before serving by using a piece of bread to gently blot up any oil from the top.

Serve hot with flatbread, crackers or warm Honey Oatmeal Bread.

Makes enough for twenty.

Minestrone Soup

MINESTRONE SOUP

Only one person has ever called my minestrone "lasagna soup" to my face—no one has ever complained. I always make more dinner than needed in case someone shows up unexpectedly, but also to ensure that I'm left with the seeds of another meal. This soup has its beginnings in leftover lasagna. Mine is made with a robust meat sauce and plenty of cheese, but any lasagna recipe will do. Leftover spaghetti translates into hearty minestrone as well, but it's a little more difficult to disguise.

1 cup of dried great northern or some other white bean //
1 bay leaf // olive oil // 6 cloves of garlic // 2 medium red onions //
2 celery stalks // 4 small carrots // 2 potatoes //
1 tablespoon each of dried oregano and dried basil //
2 large scoops of leftover lasagna //
1 can plum tomatoes (28 ounces) // 2 small zucchini //
¾ small cabbage (about 3 cups shredded) // 7 cups of water //
salt and pepper to taste

Have the beans soaked and rinsed. (See page 123.) Put them in a mid-sized pot with the bay leaf and cover with cold water. Bring to a boil, then reduce the heat and simmer covered until the beans are soft, about an hour or so—longer if the beans haven't been pre-soaked. Check on them from time to time to skim off any accumulation of foam. Drain and set aside, reserving a cup of the cooking liquid.

While the beans are cooking, prepare all the other ingredients for the soup: mince the garlic and the onions; chop the celery; peel and chop the potatoes and carrots (for variation try an oblique cut on the carrots); quarter and chop the zucchini into good-sized chunks; and shred the cabbage.

Separating the layers of the lasagna is a little messy but necessary: Scrape the filling layers into a small bowl and cut the pasta into strips or squares. Finally, break the plum tomatoes into small pieces.

When everything is ready to go, place a large Dutch oven or heavy-bottomed pot over medium-high heat and pour in enough olive oil to generously coat the bottom. Add the garlic, onion and celery and cook gently until soft. Stir in the carrots, potatoes and herbs and cook partially covered over low heat until tender, 10–15 minutes.

Push the vegetables to one side of the pot to make some room. Stir in the filling salvaged from your lasagna, and bring the heat back up to medium-high, adding a drizzle of olive oil as needed to keep it from sticking to the bottom of the pot. As the filling begins to sizzle draw all the components in the pot together, then pour in the reserved liquid from the beans.

Add the cabbage and zucchini, cover the pot and cook gently for 2–3 minutes. Stir in the tomatoes, beans and pasta, and top up the soup with 7 cups of water. As you are not using stock, you'll likely want to add a good amount of salt and pepper at this point. (How the lasagna is seasoned will also have some bearing.)

Bring just to a boil, reduce the heat and cook gently so that all the flavours come together nicely, about another 15 minutes. Make adjustments to the thickness of the soup by adding water as you see fit. Serve with Parmesan and tear in some fresh parsley, oregano or basil if you have it.

Makes enough for ten to twelve people.

POLLO RELLENO

A pan full of a half-dozen roasted chickens being flattened under a stack of thick cutting boards and heavy pots was so arresting that I eventually forgot I'd come to ask the Artigas cook for an asado recipe. Henry claimed Pollo Relleno is not particularly unique to Uruguay. I think the stuffing made of cheese, ham, red peppers, raisins, carrots and hard-cooked eggs makes it distinctly South American. That day he served it to Uruguayan dignitaries. It's a special meal.

A good-sized whole boned chicken (3 to 4 pounds) // 3 eggs //
2 carrots // 1 sweet red pepper // 1 pound smoked cooked ham // 1 pound mozzarella //
½ cup raisins // 3 cloves of garlic // 2 tablespoons grainy Dijon mustard //
1 teaspoon dried oregano // olive oil // dried sage and thyme //
salt and ground pepper // darning needle and thread

Wash the chicken and pat dry. Henry's chickens were completely boned, but I like to keep the drumsticks and wings intact. Have the eggs hard-cooked, cooled and peeled. Peel the carrots and core the pepper, then cut them into long thin strips. Dice the ham and mozzarella and toss them together with the raisins in a bowl.

Open the chicken and lay it flat, flesh side up, on your work surface. Mince the garlic and combine it in a small bowl with the mustard and oregano. Use your hands to rub the paste all over the inside to make a nice foundation for the filling. Scatter a layer of the ham, cheese and raisin mixture overtop followed by a spray of peppers and carrots arranged so that the tops and tails are at either end of the cavity. Next line the hard-cooked eggs down the centre and alternate layers of the remaining ingredients.

When you have used up all the components, gently pull the sides of the chicken together, coaxing and jostling all the filling into the cavity. Sew a seam down the length and both ends of the chicken. Drizzle a little olive oil into a roasting pan and put in the chicken, seam side up. Rub some of the olive oil into the skin and season liberally. Roast in a 350°F oven until crisp and golden and the chicken's juices run clear, 1½–2 hours.

Transfer the chicken to a rimmed baking sheet or platter. When the chicken has cooled slightly, place a large cutting board onto the chicken and weight the board with bricks, a heavy pot or books. Let cool to room temperature, and then remove the weights, cover with plastic wrap and refrigerate. Carve into slightly slanting crosswise slices. Henry serves it chilled: I prefer it at room temperature. Pollo Relleno is lovely con ensaladas.

Makes enough for eight people.

CRANBERRY FOOL

The Russians bring enough provisions for a year from home. Cranberries are an obvious choice because they keep well and are a decent source of vitamin C. A gallon jar of cranberries was my muse for this surprisingly luxurious combination.

1½ cups sugar // 1½ cups water // 3 cups cranberries // 2½ cups whipping cream //
1 tablespoon sugar // ¼ cup Grand Marnier

Combine the 1½ cups of sugar and water in a mid-sized pot and bring to a boil over medium-high heat, stirring until the sugar dissolves. Add the cranberries and return to a boil briefly, then lower to medium heat and cook, stirring occasionally, until the cranberries have popped, about 10 minutes. Pour into a small bowl, cover and refrigerate.

When the cranberries have completely cooled, whip the cream in a large bowl until it starts to thicken. Beat in the remaining tablespoon of sugar. Add the Grand Marnier and continue beating until the cream holds stiff peaks.

Gently fold 1½ cups of the cranberries into the cream. Alternate layers of cream and reserved cranberries for seams of colour and texture in each glass or bowl. Chill for half an hour before serving.

Makes enough for six to eight people.

CHOCOLATE CAKE TWO WAYS

Recipes that require few fresh ingredients are handy even if you aren't cooking in far-off places. I served this cake dusted with confectioner's sugar or topped with a thick chocolate butter icing countless times in tree-planting camps on the Canadian Shield. In Antarctica an abundance of whipping cream, a large chunk of chocolate and a gift of Christmas spirits were the inspiration for a pairing that became my signature dessert: Chocolate Cake with Grand Marnier Chocolate Ganache. I tell people this cake doubled as currency for helicopter excursions. In truth the Uruguayan pilots were so hospitable I imagine they would have invited us along even without an offering of sweets.

For the cake
2 cups sugar // 3 cups all-purpose flour //
2 teaspoons each of baking powder and baking soda // 1 teaspoon salt //
6 tablespoons cocoa powder // ½ cup butter // 2 cups hot water //
2 tablespoons vinegar // 1 tablespoon vanilla

Preheat the oven to 350°F and grease two 8-inch round cake pans.

Whisk all the dry ingredients together in a large bowl and set aside. Cut the butter into small squares and place in a medium bowl. Pour in the hot water and stir. When the butter has melted, add the vinegar and vanilla and continue stirring until the mixture has cooled slightly. Add the liquid ingredients to the dry and mix well.

Scrape the batter into the cake pans and bake in the centre of the oven. The cakes will be ready when their tops spring back when lightly touched or when a skewer inserted in the centre comes out clean, about 25 minutes. Set the cakes on wire racks and let them cool completely in their pans before you turn them out to ice.

While you are waiting for them to cool make your butter icing or ganache.

For the chocolate butter icing
½ cup butter at room temperature // 4 cups icing sugar // 4 to 6 tablespoons milk or water //
5 tablespoons cocoa powder // 2 teaspoons vanilla

Using an electric mixer, cream the butter with half of the icing sugar until the sugar is evenly distributed, then stir in 2 tablespoons of milk or water. Sift the cocoa into the butter and mix well. Add the vanilla, followed by the remaining icing sugar. Beat until smooth. Add enough of the remaining milk or water to make a thick, creamy icing.

This recipe makes a generous amount of icing even by my standards. If you halve each layer horizontally to make four layers you'll still have enough.

For the chocolate ganache with Grand Marnier
16 ounces bittersweet or semisweet chocolate // 3 to 4 tablespoons Grand Marnier //
2 cups whipping cream

Chop the chocolate into small pieces and combine it with the Grand Marnier in a medium bowl. Pour the cream into a heavy-bottomed pot over medium heat and warm it until it begins to steam and a few bubbles appear on the surface. Remove it from the heat immediately and pour it over the chocolate. Stir gently with a wire whisk to make sure that all the chocolate has come in contact with the hot cream; let it rest for a few minutes then stir slowly until smooth.

To glaze your cakes, place the layers side by side on a wire rack set on a rimmed baking sheet so that any puddles of ganache can be reclaimed. While the ganache is still quite hot, spoon a small amount onto the centre of each layer and use an offset spatula to spread a thin coating across the tops and around the sides of the cakes, then place them in the freezer for 5–10 minutes. This is called the crumb coat because it prevents little particles of cake from messing the surface of your icing. As the crumb coat sets, stir the ganache often so that it cools evenly— you want it to be barely warm, but still fluid enough to pour.

You're going to make about three coats with the ganache. Remove the layers from the freezer. Pour an equal amount of ganache onto the centre of each; smooth it out towards the edges and around the sides using a spatula. Set the layers aside in a cool place to let the glaze firm up a bit and repeat this process until you have used up most of the ganache.

To reclaim the ganache, gently scrape any chocolate that may have dripped onto the baking sheet back into the bowl. Place over a pot of barely simmering water and stir until just melted. Continue using to coat the cakes.

When you have created a nice thick, even surface on each cake, settle one on top of the other on a plate and use any remaining ganache to smooth the seam and finish the top and sides of the cake to your liking.

For a really decadent cake, spread marmalade on top of the bottom layer before the crumb coat or finish with chocolate curls or shavings.

Radio Sasha, when asked if the cake was good: "Yes, it was good. It was big."

Makes enough for ten to twelve people, or two pilots or one radio operator.

"Sunday April 13, 1947

We were not able to go ashore today because the ice was packed into the beach. Finn had a marvelous flight…[McClary] and Woody broke into some of the E rations…Anywhere they can make trouble they do.

The two Bobs on Neny Island seem to be getting along fine. Jennie has learned to make water with the evaporator. Jorge thinks he is in love with her and follows her around everywhere until now people are beginning to talk about it."

—Jackie Ronne, Ronne Antarctic Research Expedition, Stonington Island, 1947. She and Jennie Darlington were the first American women to overwinter in Antarctica

Washing up after dinner at the winter quarters, Cape Denison, Frank Hurley. 1911–1914

EQUALITY LIBERTY INFINITY
Carol
———

Deeply regret delay only just managed to reach hut effects gone but lost my hair you are free to consider your contract.

—Douglas Mawson telegram to his fiancé Paquita Delprat

Deeply thankful you are safe warmest welcome awaiting your hairless return regarding contract same as ever only more so.

—Delprat's reply telegram to Mawson

Antarctic explorer, scientist and conservationist, Sir Douglas Mawson's 1911–1913 journey across unexplored coastal regions was "probably the greatest story of lone survival." Mawson's Australasian Antarctic Expedition party celebrated Midwinter Dinner on June 22, 1912, eating noisettes of seal and awaiting the sun's return. They faced 300 km per hour winds. Mawson, Xavier Mertz and Lt. B.E.S Ninnis left the group to explore far eastern Antarctica. Ninnis disappeared into a deep crevasse with a sledge of supplies and Mertz likely perished of hypervitaminosis from eating dog liver. Mawson miraculously returned, near-death and unrecognizable. He had left Australia December 11, 1911, expecting to see his fiancé 15 months later. Mawson returned in February 1914 and soon after married Delprat.

He wrote *The Home of The Blizzard* after his return. Mawson was an early advocate for Antarctic research stations. He also left an extreme Antarctic menu plan including dog and skua.

Antarctica transformed me. I became braver and certain I wanted to work on global justice. I felt fortunate to experience something grand that made me feel cosmically tiny.

Soon after Antarctica I went to Africa's "first world war" in the Congo (then Zaire) with a medical humanitarian group. I also worked in Southern Sudan where people were dying from a famine caused by political action, not only drought or a lack of food. I met people who had coped, survived and even hosted a community dinner to celebrate an impending marriage. But I saw too many freshly dug graves. Never before have I wanted to stay in and leave a place simultaneously. I want to be a witness of our times.

I encountered a South African woman caring for her deceased children's children in a community ravaged by HIV/AIDS. Documenting how her community grieved and responded to the crisis, she said, "If you don't write it down, maybe it didn't happen."

Antarctic explorers had to expect suffering as they chose to go to an inhospitable place. There as everywhere we are at nature's mercy and are reliant on one another. We are also resilient. The world isn't divided neatly into humanitarians and those who exploit. As I felt in Antarctica, I also felt when working with some of the world's most vulnerable peoples that we have to do more than 'do no harm.'

GETTING THE COLOURS DOWN

FEBRUARY 16–29, 1996

It's the same at all remote work sites—as if you belong to everyone.
Everyone and no one.

FEBRUARY 16, 1996

WT   11:30 p.m. Rain and wind all day long and still now. Leaky windows. Haven't had
weather this bad since January; still, it doesn't seem to bother this team. Sean says the couple
from California wants to stay for another camp. If only they'd been here a month ago. You
have to admire their dedication, but it would mean working on the cleanup through our days
off. Lena turned green when Sean broke the news and I can tell he's completely knackered
by the way he vented to me, "They are not my friends; they are my job." The things you think,
but never say. Hope the phones are down at Frei and they can't switch their flights. Sean has
to shift his focus to the logistics of the loadout: ask the Chileans if we can use their front-end
loader; confirm with the captain we've bound the fuel pipe properly, and we all need to experience more of King George.

   Busy again, though managed a visit in Diesel with Sasha and Volodya while doing laundry.
They were especially animated today, as if to apologize for all that silliness at the beginning
of the month, as if to say they missed me. Finally got most of the soot out of my aprons and
oven mitts and made a sauna appointment for tomorrow at 4:00. Hope to get my own laundry
done then. Must remember to give Lena something for the Korean fête and have her arrange
a recipe exchange. Wish I could go—feel a little like Cinderella.

FEBRUARY 17, 1996

Cloudy, but pleasant. Walked to Stoney Bay to deliver treats to the volunteers. They were
laughing and joking. I can't believe the progress they made around camp—$5\frac{1}{2}$ barrels of

mixed waste. That's more than twice the amount any other team has collected. Rain suits. If only all the other teams had brought rain suits.

Hiked new way home way up high in the hills above camp. Trying to get everything from this place—wanting to stay—the idea of overwintering is tempting. Maybe I'll camp at Stoney one night on the next break, take in all that emptiness? In the end Lena got me a seat for the ride to the Korean base, but I was in the middle of baking and missed the flight. Zuniga invited Lena and me to Artigas for a party tonight—made it clear we won't be the only women. A psychologist, a meteorologist, a chemist and an architect arrived on the last plane.

Late, late.

What a shaker! Singing; guys jamming on guitars; beating on overturned 50-gallon barrels; dancing. Even with the new arrivals from Uruguay we're token females, but it's safe and we immediately know it, feel respected and somehow accepted. It's the same at all remote work sites—as if you belong to everyone. Everyone and no one. The idea of someone at home or not.

2:00 a.m. Chilean time/3:00 a.m. Uruguayan time. Party still on. So dark now—a month ago this would have been first light. Classical Spanish guitar serenade from the guy riding in the back-seat on the drive home. From up in the hills Frei's twinkling lights make the Chilean base seem more like a quaint village and I understand why Zuniga likes it. It must bustle when all their wives and children are there. Funny man. He acts as if there is something going on between us especially tonight with the Uruguayans—reaching for my hand, cutting in on the dance floor, and playing with my hair. I don't have a clue where it's coming from. All innuendo, best to let it pass, let everyone keep their dignity.

FEBRUARY 18, 1996

Exhausted today. January weather has returned. Constant rain and wind. So hard to walk up the hill. Great debrief with the group of volunteers. They've been a real boost for morale with the added benefit of finally having a geologist on the team, Jeffrey. Think we all learned something from him on our hikes and now I know the names of all my rocks. Wish I'd asked him what makes some jasper red and some green. Some talent on the football pitch as well at our match in the Chilean gym. Uruguay took Bellingshausen 22–2, but good to see men in shorts. Dinner afterwards at Canada House: Quique, Jose, from Uruguay and the VIEW team—Ilya, Sad Vlad, Dima, and Hilltop Sasha. Party fun with guitar-playing and one of the volunteers on the harmonica. Up very late again.

FEBRUARY 19, 1996

4:00 a.m. wakeup for 5:00 a.m. pickup by *Petrov*. Ship late. Tolerant but tired volunteers. Mist and rain. Spent lounging, breaking KGI rules with Sean. Poquito sleep. Mail finally! Lots of writing, relaxing and reading the magazines Carol sent. Ah the *Vogue* spring collection and *The New Yorker*. Wish Carol could swing coming for the last camp; think she'd be proud of our progress. Early night at last and good sleep until Quique let himself in to return my flashlight. Startled, not scared.

Whale bones, the Drake side of King George Island, 1996

The Drake side, 1996

Camp 14 Debris Collection (Feb. 14–19)
Stoney Bay Area 13: 1 barrel mixed
Bellingshausen Area 8–10 and 12a: $5\frac{1}{2}$ barrels mixed

> I've had a lifelong romantic attachment to the Antarctic as well as an interest in the scientific work. This project provided me with a real physical experience of the life of dedicated people working in unusual conditions.
>
> —Volunteer Geoffrey, South Africa

FEBRUARY 20, 1996

Laundry pretty much the whole day after leisurely morning. Visit from Quique to apologize for frightening me in the middle of the night—odd choice to drop by with a load of jeering soldiers in the truck bed. Clean up Canada House. Laundry. Most satisfying. I think my clothes are finally clean and the bathtub is at last making bubbles.

Talk with Volodya Driver at Diesel was really unsettling. Insisting that women at stations are a problem, saying Lena and I are a problem. Wow; he really knows how to cut through when he wants to—so offensive, almost hurtful. I think we know how to behave and have done so. And then he went on to tell a story about the Arctic where there was one woman on the base. A spurned lover killed himself and two other men got in a gunfight over her. Glad Sean was with me. Told V if I'm a problem, I don't have to visit. He claims I'm not a problem to him anymore because he took his heart in his hand, that he was in the Arctic for six years before coming to Bellingshausen and is more accustomed to being alone than some of the others. Had to resist reminding him that he swung the first punch earlier this month. Assured by Sean and radio hut guys afterwards we are not a problem. Sad Vlad and Lena up to Canada House after dinner. Shared last bottle of wine as I made hairclips. I like the one made from rusty bits best.

> Patience Camp, 1914
> "The food now is pretty well all meat," Greenstreet wrote. "Seal steaks, stewed seal, penguin steaks, stewed penguin, penguin liver…The cocoa has been finished for some time and the tea is very nearly done."
>
> —Caroline Alexander, *The Endurance: Shackleton's Legendary Antarctic Expedition*, 1998

The more laundry you have, the more time you have for tea—On laundry days I lost myself in a whole cleaning ritual. I'd line up basins and pails to soak my oven mitts, tea towels, tablecloths, and aprons while I was in the sauna and drop them in the washing machine during my bath. If there was lots of laundry I might hit the sauna and stream twice before running anything through the wringer.

Volodya Driver always let me hang laundry by the generator fans and sometimes he helped. In between loads we'd have tea in the watchman's room. There was always someone willing to put on the kettle and clear the chair beside the desk. A gang would often appear as if they'd been tipped off, and snacks from the kitchen would follow. All those men huddling close—that room became our hearth.

Sasha Diesel serving tea in the watchman's room, Bellingshausen, 1996

I have records of the many conversations I had with Volodya. Quick sketches on scrap paper: a tent by the side of a river to show me he liked camping ("kemping"); a woodland to show me he liked foraging for wild mushrooms ("lesnyje gryby"); a guy with a fishing pole to show me he liked fishing ("ribalka") and drawings of a woman (me) with children clinging to her. Some days my heart would break the way he opened his for mine. I'd volley back a light and bright drawing of a woman paddling solo and wondered if he thought I was a bit dim the way I just didn't understand. I hope he never thought I was heartless.

Sasha Diesel made the best tea. He spoke less English than I did Russian so we'd default to Spanish, which was equally dubious. Mostly we'd sit in companionable silence making things. Jewellery-making started with the first quartz crystal. I was so touched I started to fidget, turning it between my fingers. After a while I spied a paper clip on his desk and twisted it around the stone and up into an earring hook—this set the tone for all of our visits. With every new trinket I made he'd up the ante with another stone, Russian coins or a finer-gauge wire to work with. I started to bring rusty bits I'd found; he'd search through drawers to find another offering. In time he started calling me "soroka" (magpie).

One day between camps, between laundry loads, between cups of tea he motioned for me to follow him. I forget the way—somewhere through the main part of Diesel, past the sauna, to a machinists' shop I didn't know existed. Wordlessly he ground a tool for me to punch holes, and showed me a metal press. We found two copper washers and pressed them until they were gleaming with undulating edges. I tapped a pattern and holes for hooks with the punch and added these new dangly bits to the earrings I was wearing.

There was a buzz in the mess I couldn't get a read on at lunch that day until I heard chairs shuffling and noticed a table of men craning their necks to look at me, hands to their ears so I'd know why. And then a chorus of "molodets." I think it means "clever" or "well done" or "good man." I beamed and regretted ever having to learn how to say "Ne cejchas, nyet vremeni, ya rabotayu." ("Not now, no time, I'm working.")

FEBRUARY 21, 1996

5:30. Having a lazy afternoon. I spent the morning in Chile touring their kitchen, mess and stores, chatting about food and recipes with Fernando and Manuel. The way they pointed out wares on the tour of provisions and bodega was comical—as though this might be the first time I'd laid eyes on such products. Given shampoo, a demitasse, and some mystery sweet thing Fernando's going to teach me how to cook. Promise that he is going to bring me cognac and some other surprises when he comes on Sunday. Manuel said something to the effect of "You are like Princess Diana." Must be my fashion sense.

Lunch afterwards with Zuniga and officers, as well as officers and captain from Chilean supply ship that's downloading oil today. They served cazuela as it is Wednesday and the reason for my visit. Very good, but I like mine better. Got a few more recipes from Zuniga — hope Fernando will be able to translate for me. Z asked me to make cazuela for him before he leaves. Must remember to make bread with less honey when he comes, perhaps deliver a loaf to him and guys in the kitchen. Promise from Felipe he will take us flying to other stations when the weather is better.

Other events: saw two crabeaters migrating through camp today. Volodya Driver taught me the Russian for old maid ("staraja deva") Ouch! Saw Su-Chong from Great Wall—invited for dinner there on the 23rd. I'll go early to hang out in kitchen. Plans to have Diesel guys for dinner tomorrow. Pancakes tonight—my favourite. Get recipe from Vladimir Cook.

Feeling crabby tonight—tired of being pursued and tired of being polite. Just tired. Really feeling need for personal space. A walk home in the wind helped and clear skies too. First time I've really noticed the stars. I don't know the sky down here.

FEBRUARY 22, 1996

I want to make sure I remember:

The way the mist comes down right upon us

Twinkling lights on the ships in the harbour

Sunlight on Argentinean and Korean bases across the bay

Silhouette of Flat Top and the big rock at Stoney Bay

Tinkling of ice as it breaks up on my lake, ice sculpture brash bits at low tide

Penguins running with outstretched wings

Midnight sunshine

Cod-jigging in January and my first taste of fish in Diesel

Laundry day—wringing clothes, like Nana

December and January mud

Our elation when we could finally put our rubber boots away

Wind pushing me down the mountain—wind pushing me up the mountain

Ladders, my rust collection and cool old things

Waking up at 3:00 a.m. to see the sun rising

Warming water on the stove—a basin, soap, warm washcloth—all the comforts

Moss changing colour, becoming so thick that it looks like bushes and rustles in the wind

Seams of green and red jasper in rock

Sitting on water tank on hill overlooking camp

Jewellery making in Diesel, tea at radio hut

Sergey's evacuation and return, and the isolation and helplessness we all felt

Smell of the Drake on a still day

Whale and penguin graveyards

Fur seals' coats, glistening in the sun, like Northern Ontario black bears

Glaring elephant seals, inquisitive penguins, crabeaters in the middle of everything

Patterns on soil made by rocks and wind

Going to shed for eggs in the morning

Clouds lifting—catching a glimpse of the Drake

Measuring the days and the weeks by iceberg flow

Wind hitting air vents and propane tank box, rain through the window onto my bed

Numbing waters of the lagoon

Drying my hair in front of the Diesel generator fan

Clean, dry clothes

Calving bergs at Drake
Wondering what others will go home with
People telling me I'm like a bumblebee or a mouse, darting place to place
Lena telling me I have the hands of a small boy, fingernails always dirty
Time to think, time to play, time to wonder

Absolutely magnificent day. Sun, warm, no wind. Up early, wrote all morning, finally dressing at noon. Learned how to make cabbage pie with Lena. "It feels like home. It's my family's pie. My mother always cooks it, my sister, everyone."

Later: Skiing on rise behind Canada House. Snow bad—a thick layer of corn snow but it won't hold to the ice underneath. Grade not particularly difficult, but I'm cautious given the rocks piled where the pitch ends and the dodgy runoff so close to the best snow. Skuas overhead shadow my run and climb back up for another. And where the snow ends moss and lichen so thick I can't resist putting my face to it. Play of sun on Nelson Island glacier. Skuas approach boldly as I lie on the mountaintop. Dancing, yelling, laughing. Slide down the unskiable parts, walk home by the lake, ski poles up for protection. Dodge diving skuas. Sauna. Then up the hill to make dinner for the Diesel guys—a real pleasure to cook for them. Got some chuckles when I served chicken skewered on nails.

Chicken bites; zuch toasts and tapenade; soup; cheese; cabbage pie; roast beef with cognac sauce; baked spuds with sour cream, bacon, cheese, scallions; chocolate mousse.

FEBRUARY 23, 1996

9:00 a.m. Wake up to another gorgeous day. Don't remember much else other than shutting the curtains—a ploy on Sean's part. Bit of a lie-in today. Hard to get going. Eventually down the hill at noon for sauna and work in Diesel room. Lunch with guys putting things on my plate, insisting that I eat. Afternoon in the garage with Sasha helping me make earrings.

Walk to Great Wall station for cooking lesson and dinner. Teach Mr. Wong to make bread and he shares secrets of dumpling-making and other Chinese foods. Su Chung translates. Secret of Chinese cooking? Lots of oil, he says. Fancy dinner in conference room, taste dumplings we've made. Simple and delicious. Drive home at 9:00. Laden with gifts from their kitchen—century eggs packed in rice hulls and canned lychees.

Walk up hill. Lonely tonight. Colder—frost on Niva and ground. Happy to see fire burning in barrel outside Canada House. Am conscious of liking someone to come home to.

Two new stones from Sasha.

Let us follow the narrow sledge-tracks that the little black dots of dogs and men have drawn across the endless white surface down there in the South—like a railroad of exploration into the heart of the unknown. The wind in its everlasting flight sweeps over these tracks in the desert of snow. Soon all will be blotted out. But the rails of science are laid; our knowledge is richer than before.

—Fridtjof Nansen, Introduction to *The South Pole,* Roald Amundsen, 1912

Breadmen and women dancing, Bellingshausen, 1996

Volodya's pancakes and tea in Diesel room, 1996

FEBRUARY 24, 1996

10 a.m. Cloud, cold wind from east bringing in brash ice. Volunteers are not coming until this afternoon. 12:30 p.m. Winter blizzard with very high winds. Volunteers forced to stay on ship. Another five-day break—hope I can pack everything in. Funny how today worked out with me holding off dinner prep a little longer than usual. Did my baking, though, which I know the guys will enjoy. Dinner with the Russians—wind so strong I nearly didn't make it down the mountain. Hiking across the base, three guys standing on the porch waiting, two run out to help me—funny, I had to come all the way to Antarctica to find gentlemen.

Well, most of them.

Here I am so far away needing solitude. You'd think I could find it. The attention is overwhelming. I'm craving anonymity.

A beautiful night with sunset through clouds. Sad Vlad gave me his Soviet navy jacket and holster as a memory of Satellite House parties—it'll be a knockout in Toronto. Think I'll leave the holster in the dressup box. Visit from Fernando to deliver cookbook, Christmas tree ornament and ingredients to make mote con huesillo. It's going to take some magic to make dried peaches and pearl barley taste good. And what else of the day? Walked with Sean up to the point. Scrambled up some rocks, lay on a ledge, watched the birds overhead. Listened to the waves and the wind. Watched a snow squall pass by. Later, scree running down the mountain. Hatched a plan to go skiing on the glacier.

FEBRUARY 25, 1996

Lost myself in a letter to *Morningside.* Took minestrone, bread and Parmesan down to the guys. I'm not sure if they like my cooking or like that it's different. Doesn't matter I guess.

Talked with Volodya Driver after lunch. Questioned why such a good woman likes such dirty things. Told me he wished I could speak to him for just one hour a day. If I got it right, he advised me to have children because they can look after me when I'm old. If not, life will be hard. I responded, "That's life—c'est la vie." He replied, "Chudo! Ty prosto chudo." I think—I'm not sure—but I think it's something about being a miracle. Lena explained, and the guys in radio hut assured me it was a compliment. Hung around with them for a bit there and then up to Satellite House for our usual program: Russian folk songs with Laura Branigan dance party interludes. It will be a long time before I listen to eighties pop at home. Best part of the night was learning to sing "Milen'kij ty moj." A little action in the middle of the night when one of the ships brought in an injured man for a medevac to Punta Arenas. Makes even the healthiest of us feel vulnerable and this poor guy had no medical or travel insurance— every traveller's worst nightmare.

Behind the Curtain—When I think of the conditions Antarctic explorers endured and the eloquence of their language, not having time to write is a pretty sorry excuse. It also makes me grateful we have any record at all, and curious about the stories between the lines.

The first kiss for instance. My best guess is that it was the day Sean and I spent breaking KGI rules or when he asked me to reach over him to shut the curtains, although that seems like it could

Dumpling making at Great Wall Station, 1996

The rise behind Canada House,
a glimpse of Stoney Bay in the distance, 1996

have been something else. I think Lena knew by the way she said out of the blue one day, "Sean has a good face," looking at mine for telltale signs. I gently nodded yes, added he had nice eyes and kept cooking—it was easier that way.

But why the cryptic emotional shorthand in my journal?

I wasn't concerned anyone would read it and there weren't really any KGI rules. I wonder sometimes if I wrote around Sean because I was disappointed I had broken my own rule: Don't play favourites. In the bush that tenet translated into equal portions for everyone, especially dessert. On KGI I amended it slightly to: what you do with one person you have to be prepared to do with everyone. I kept it to dancing and tea.

FEBRUARY 26, 1996

Cloudy, cooler today, with not too much wind, but we know autumn is upon us; the ice in the boot-washing barrels and puddles doesn't melt anymore. Brash ice on the shore. Surprised that no bergs brought in by storm. An inside and clean-the-fridge kind of day. Started the cazuela for tomorrow's dinner, got a head start on pizza toppings for lunch and made cabbage rolls. Lots of writing. I feel like I'm finally processing things. But how will I share it, where will I start? Eyes and mind are tired. I'm loving this quiet. Gotta find me a mountain shack somewhere to hide in. Make things, draw, paint, think great thoughts.

FEBRUARY 27, 1996

8:40 a.m. Left the blackout curtains open to see first light. I can't remember whether it was 4:30 or 5:00. Perfect dawn again, this time with a sugar coating of snow and frost on the windows, almost all gone now. It's a beautiful clear day, though wind from west and clouds hovering over the Drake indicate change. Hope we get some sun again later as I have a full morning ahead of cooking. What a treat to cook when there's no real pressure.

Stood up by two nations at dinner, but had a very nice time with the Uruguayans. Yummy dinner. Doc Roberto says that Uruguayan cazuela is thicker with mushy rice—my cazuela seems more like soup to them. Fernando arrived in time to try my mote con huesillos, but was really uncomfortable joining us at the table. He told me later that cooks can't eat with officers, which might explain why Mr. Wong and Su Chung were no-shows. Great yerba mate cup from Roberto, recipe from Jorge for humitas and promise he'll make me something on the second of March, also promised a helicopter ride. These guys are gentlemen as they invited Sean along. Too bad he's not a woman. Then again.

FEBRUARY 28, 1996

Sun, cloud, sun. Sunset. Rain and a bit of a blow for the night; antenna clicking in the wind. Packing and planning today. Potluck meals down the mountain. Lunch: Took pasta puttanesca and mote con huesillos. Dinner: Took cazuela, bread, cheese and Russian vinaigrette. Squirrelled away some vinaigrette for Volodya Cook for official taste test. Pancake-making lesson with him—remember to use hands to form them. Get recipe.

Dietary and aesthetic research/taste-testing with Lena, Hilltop and Radio Sashas, Dima and Ilya: biscotti for Hilltop Sasha, chocolate chip cookies and ice cream for the others.

Pathway from Canada House to main Bellingshausen base, 1996

FEBRUARY 29, 1996

Goodbyes began today. Up at 6:00 to get a shot of Radio Sasha at radio call. He posed for me in front of the wall of radios and computers. Over to Diesel to drop off juice, lemons and hot chocolate. Sasha Diesel was snoozing when I arrived, but jumped up to put the kettle on and then got the idea to have me make the wakeup calls to Volodya Cook, Volodya Driver and Vassiliy: "Dobroje utro. Govorit Wendy. Pora vstavat." ("Good morning. It's me Wendy. Time to get up.") Russian men laughing is a fine, fine sound.

Sasha Diesel poured tea and we sat making paper airplanes for a time until he signalled for me to follow him to the machinists' shop. He ducked behind equipment and came out with a crate. With his back to me he reached into it and briskly put something heavy, covered in a dirty chamois in my hands. I recognized it at once—that exquisite chunk of petrified wood the Chinese scientists found on our first Drake trip—a gift for me to take home. I can't. It still belongs here. His plan B was right there: material to patch my jeans, and a colander like the one in the abandoned buildings we explored last month. Oh my heart.

Arrived in mess at 8:15. Breakfast finished, though guys were still gathered. Moved by Vadim's gesture—he has the kindest eyes—and the way he brought me the last wedge of cheese, his entourage translating that he wanted to make sure I got a piece. Learned word for cheese ("syr") from three teachers—or was it four?—all looking at me as if they were teaching a small child. Is this my childhood?

Camp 16 Debris Collection (Feb. 24–29) Stoney Bay waste and sorting piping
Bellingshausen Area 12: ½ barrel glass
Bellingshausen Area 2: ½ barrel mixed

CHICKEN TERIYAKI BOUCHÉES

Sasha teased I was like a "soroka" (magpie) the way I was drawn to shiny things. Later our dinner guests from Diesel found out why I'd asked for nails. You can use toothpicks when you make these at home.

6 boneless, skinless chicken thighs // 2 cloves of garlic //
1 thumb of fresh ginger (about 1 teaspoon minced) // ¼ cup soya sauce //
1 teaspoon rice vinegar // 2 tablespoons brown sugar // 2 tablespoons warm water //
1 green onion // sesame oil (optional)

Cut the chicken in pop-in-the-mouth size pieces and place in a medium bowl. Peel and mince the garlic and ginger and toss them together with the chicken. Whisk together the soya sauce, brown sugar and warm water until the sugar has dissolved. Pour the mixture over the chicken, turning each piece so that every surface gets to know the marinade. Cover the bowl tightly with plastic wrap and refrigerate for at least 2 hours, turning at the halfway point.

Place the chicken in an ovenproof dish (I like a small cast-iron skillet) and broil until the juices run clear and the edges of the chicken morsels are nicely browned. Garnish with thinly sliced green onions and a dash of sesame oil (if desired). To serve: pierce each piece with a toothpick (or nails, sterilized in boiling water, for authenticity) and place on a board or plate.

Makes about 36 pieces.

ZUCCHINI TOASTS WITH TAPENADE

*I may never have thought of this sweet and savory combination if I hadn't run out of crackers.
Zucchini bread cut into bite-sized pieces, toasted and topped with a lemony tapenade makes
a delicious appetizer. You can use the tail end of the loaf or make it specifically for this.
Your favourite carrot bread or muffin just might work too.*

Zucchini Spice Bread
2 medium zucchini // $\frac{1}{2}$ cup maple syrup // 6 tablespoons butter //
2 eggs // $\frac{1}{2}$ teaspoon vanilla // 1 cup all-purpose flour // 1 cup whole wheat flour //
$\frac{1}{2}$ teaspoon salt // $2\frac{1}{2}$ teaspoons baking powder // $\frac{1}{4}$ teaspoon each of nutmeg and ginger //
$\frac{1}{2}$ teaspoon each of allspice and cinnamon // $\frac{1}{2}$ cup walnuts // $\frac{1}{4}$ cup currants

Preheat the oven to 350°F and butter a medium loaf pan or a muffin tin. Grate the zucchini
(you should have about 2 cups). Place it in a colander over a bowl for 10–20 minutes, then
press out any excess moisture. Put the maple syrup into a large bowl and beat at high speed
for about 5 minutes. Melt the butter and add it to the bowl along with the eggs and vanilla,
then beat several minutes more. Put the dry ingredients into a medium bowl and whisk until
blended. Add the dry ingredients to the maple syrup mixture in three stages, alternating
with the zucchini so that you begin and end with the dry. Blend gently but thoroughly after
each addition. Stir in the nuts and currants. Scrape the batter into the loaf pan or muffin tin.
Bake until golden brown, 55–65 minutes for the loaf, 18–20 minutes for muffins or 10–12 for
mini-muffins.

Makes one loaf, 1 dozen muffins or 4 dozen mini muffins.

Tapenade
1 cup kalamata olives // 4 cloves of garlic // 1 tablespoon capers // 2 to 3 anchovies //
2 tablespoons olive oil // juice of half a lemon // 1 tablespoon lemon zest

Pit the olives and combine them with the garlic, capers and anchovies in a food processor
or blender. Pulse to form a paste. With the motor still running, drizzle in the olive oil and
lemon juice until blended. Stir in the lemon zest.

Makes about $\frac{3}{4}$ cup.

To make zucchini toasts, slice the zucchini spice bread or muffins into manageable pieces—
about $\frac{1}{4}$ inch thick and two bites worth in size. Lay the slices on a baking sheet and broil each
side until the edges start to brown, about 2 minutes.

Lena's Cabbage Pie

мыс б.н. 0,5 км

7.6   7.8

1:2 000

В 1 сантиметре 20 метров

лошные горизонтали проведены через 1 метр

Мензульная

Северо-западн
производственн

LENA'S CABBAGE PIE

Cabbage is right up there with root vegetables, dried fruits and legumes for reliable expedition foods. My appreciation for the humble cabbage's versatility increased immensely over the duration of our project, especially after I made this savoury pie with Lena. The pastry is an exceptional hybrid of a yeast dough worked like puff pastry. Served with a light salad, a good sharp cheese, dried sausage, pickled onions, cornichons and grainy mustard it makes an impressive lunch. Lena claims it is her family's pie—it could be your family's too.

For the pastry
6 eggs // 1 cup warm milk // 1 tablespoon yeast // 2 tablespoons sugar //
3 cups all-purpose flour // 1 teaspoon salt // 1 cup butter at room temperature //

For the filling
1 onion // 3 cloves of garlic // 1 medium cabbage (about 8 cups chopped) //
3 bay leaves // 1 teaspoon dried basil // 4 tablespoons olive oil //
salt and pepper to taste // a darning needle

Hard-cook four of the eggs. You can keep the yolks nice and yellow if you start them in cold water; cover and bring to a boil then remove the pot from the heat immediately. Let them stand for 20 minutes and then drain and submerge the eggs in cold water until cool enough to touch. I like to remove the shells right away so that I'm not in a rush at the last minute.

To begin the dough, dissolve 1 tablespoon of the sugar in the warm milk in a medium bowl. Whisk in the yeast and let it stand until frothy, about 10 minutes. When the yeast has proofed, whisk in one of the remaining eggs along with the salt. Gradually add the flour, using a spoon to mix. When you can no longer stir, dust your hands with flour and gather the dough into a ball. Turn it out onto lightly floured surface and knead until it is smooth and elastic, about 10 minutes. Return the dough to the bowl after greasing it with butter, turning to coat the dough all over. Cover the bowl with a tea towel and put it in a warm place to rise for about 1¼ hours.

While the dough is rising, mince the garlic and finely chop the onion and cabbage for the filling. Heat the oil in a large skillet set over medium-low heat and cook the garlic and onions until softened. Stir in the remaining tablespoon of sugar and the basil. Add the cabbage and bay leaves and sauté covered until tender, stirring frequently so as not to brown the cabbage, about 10 minutes. Add salt and pepper to taste, remove the bay leaves and set aside to cool.

When the dough has doubled in size, turn it out onto lightly floured surface and let it sit for a few minutes. Roll it out into 9 by 18-inch rectangle, about ¼-inch thick. Spread two-thirds of the rectangle of dough with one-third of the butter (⅓ cup) and fold the unbuttered portion back onto a buttered part of equal size. Fold the remaining piece back over to make three neat layers. Wrap in plastic and refrigerate for 45 minutes (or 20 minutes in the freezer). Return the dough to the floured surface and roll, butter, fold, wrap and refrigerate again. You'll do this a total of three times. After the final stint in refrigerator, cut the dough in half and shape each piece into a disk. Wrap with plastic and refrigerate for 20 minutes.

To assemble the pie

Roll out one of the disks into a 13-inch circle on a lightly floured surface and place it into a 9-inch pie plate. Trim the edge to leave a ¼-inch overhang and roll out the remaining disk into the same-size circle so that it is ready. Chop the hard-cooked eggs and arrange them evenly in the pie shell followed by the cabbage mixture mounded up nicely. Place the second piece of dough over the filling and trim it so that it extends just beyond the bottom piece of dough. Fold the overhang under the pastry rim and flute the edge to seal. To let steam escape during baking, cut decorative vents or prick the pastry with a darning needle or fork.

Whisk the remaining egg with one teaspoon of cold water until smooth and brush it evenly over the pie. Bake in the bottom third of a 400°F oven for 15 minutes, then reduce the heat to 350°F and bake until the crust is golden brown, 20–25 minutes longer.

Makes enough for six to eight people.

CHOCOLATE AND CREAM

Many chocolate mousse recipes incorporate eggs, but you can still create a satisfying dessert when you need to work around food allergies or save eggs for breakfast and baking.

¼ cup Grand Marnier // 1 tablespoon sugar //
8 ounces bittersweet chocolate // 2 cups whipping cream

Pour the Grand Marnier into a tempered glass or stainless steel bowl and sprinkle in the sugar, stirring to dissolve. Put about 2 inches of water in a mid-sized pot and bring it to a gentle simmer. While the water is coming up to temperature, chop the chocolate and add it to the bowl. Set the bowl over the pot and reduce the heat so that the water is barely simmering. Stir occasionally until the mixture is glistening and not quite smooth. Remove the bowl from the heat and set aside. The residual heat from the bowl will finish the melting while you prepare the cream.

In a large bowl whip the cream until it forms soft peaks. Add about a quarter of the whipped cream to the chocolate and whisk together. This will prevent the melted chocolate from seizing when it hits the cream and make it easier to bring the mousse together uniformly. Gently pour the chocolate mixture into the whipped cream and use quick, delicate strokes to fold until smooth and luxurious. Transfer to a serving bowl or individual bowls or glasses and refrigerate until firm. Chocolate cream can be made a day or two in advance. Serve with chocolate shavings or curls.

Makes enough for six to eight people.

To make chocolate shavings, draw a sharp vegetable peeler down the side of a block of chocolate.

For curls, melt a small amount of chocolate over hot water and use an offset spatula to spread it over the bottom of a baking sheet. Place it in the refrigerator to set, about 15 minutes. When the chocolate has cooled sufficiently, pull the side of a spoon along it in one long definite stroke—a 4-inch pull makes a nice curl. Place the curls on a wax paper-lined pan and refrigerate until you are ready to use them.

GREAT WALL DUMPLINGS (JIAOZI)

It must have been a slow science day when I went to the Great Wall station to learn how to make dumplings. A steady stream of people abandoned their labs and crowded into the kitchen to catch a glimpse and take photographs of a foreigner getting a cooking lesson. Many offered pointers. Dumplings are a delightfully serious undertaking.

This recipe makes 24 dumplings with meat or vegetable filling. You can also halve the filling ingredients and make 12 of each kind.

For the dough
1 cup all-purpose flour ∥ ½ cup water

Mix the flour with the water in a small bowl. Mr. Wong says to use warm water in the winter and cold in the spring, summer and fall. If the water is too hot the dough will be too difficult to work with. Su Chung says his family always uses cold—everyone has their own approach. Let the dough sit for 10 minutes or so, but not longer than 30 minutes or it will be too soft to work with. While it rests you'll have just enough time to prepare the filling.

For meat filling
2 green onions ∥ 2 cloves of garlic ∥
a thumb of fresh ginger (about 1 teaspoon minced) ∥
¾ pound minced pork or beef ∥ water or sesame oil

Peel and mince the onions, garlic and ginger and put them into a small bowl. Add the meat and mix together well, then add enough water or oil to make the mixture sticky.

For vegetable filling
2 green onions ∥ 2 cloves of garlic ∥ a thumb of fresh ginger (about 1 teaspoon) ∥
1 stalk celery ∥ 10 snow peas ∥ 1 carrot ∥ 10 bamboo shoots ∥ 2 water chestnuts ∥
5 Chinese black mushrooms (soaked in advance) ∥ sesame oil

You are going to want a really fine mince. Start with the green onions, garlic and ginger and set them aside in a medium bowl. Mince the celery, snow peas, carrot, bamboo shoots, water chestnuts and mushrooms and place in a sieve, pressing out any excess water. Add to the bowl, mix well and then drizzle in enough sesame oil to make the filling sticky, about 2 teaspoons. Set aside until you are ready to make the dumplings.

To make the dumpling skins, dust your hands with flour and knead the dough in the bowl for a few turns. It will still be quite soft so you will want to work quickly. Scoop it into a ball, and then transfer it to a lightly floured surface. Divide the dough into two pieces and roll out one of the pieces into a long rope, about 1-inch in diameter throughout. Cut the rope into equal-sized knobs, dust with flour and roll into small balls. Repeat with the remaining dough—you should be able to get about 24 balls. To form each skin, press one of the balls with the palm of your hand then switch to a rolling pin. Roll out from the centre in both directions to make a uniform circle, about 2½-inches in diameter. Taper the edges as you go. Alternatively, you can form the skins using the flat side of a cleaver and a forceful smack.

Have a small bowl of water ready. Place 1 heaping teaspoon of filling in the centre of a skin. Brush water around the outer edge and fold one side over to make a half-moon. Press the edges together then pick up the dumpling. Using the palm of your hand to support the bottom, crimp the outer edges together like a fan. As you get better at it you may be able to increase the amount of filling you add and make a really fancy edge. Continue until all the skins are filled.

To cook the dumplings, bring 2 quarts of salted water to boil in a large pot or wok. Add the dumplings, return to a boil, then pour in ¼ cup cold water to take the boil down. You are going to do this two more times. When the dumplings have come to boil a total of three times, strain and rinse under cold water to keep them from sticking. Serve with malt vinegar or soya sauce.

Makes enough for four to six people.

UKRAINIAN CABBAGE ROLLS

Thank goodness for Patricia, the dear Ukrainian woman who helped look after my family when I was growing up. She never actually taught me how to make her cabbage rolls, but when I felt compelled to give them a try in Antarctica I found I could because I'd watched her so many times. There is a strong argument for letting kids get underfoot in the kitchen in here.

It was my mother's idea to top them with pineapple. The sweetness was a surprise to the Russians when I shared a batch with them—they claimed Russian food was not as "flavourful."

2 heads green cabbage // 4 onions // 6 cloves of garlic // 3 tablespoons butter // ¾ cup long grain rice // 2 cups water // 2 pounds lean ground beef // salt // pepper // 1 can tomato juice (48 ounces) // 1 can crushed or chopped pineapple (14 ounces)

To prepare for blanching the cabbage, put a large pot of salted water on to boil. You are also going to want to have large bowl full of ice water ready (the sink is even better) and some tea towels laid out to drain the leaves.

Core the cabbages and remove any tired-looking outer leaves while you are waiting for the water to boil. Blanch the cabbages one at a time for 5–8 minutes until the leaves are softened. Use a slotted spoon to remove the cabbage from the pot and then immerse in the ice water.

When the cabbage is cool enough to touch, working from the core end, gently peel back the leaves and place on the tea towels to drain. Pop the cabbage back into the boiling water for 2 3 minutes when the leaves start to resist. You should be able to pull off 12 15 leaves before they become too difficult or small to remove. Reserve some leaves to cover the rolls in the pan as well.

Repeat with the second cabbage. Before moving on to the filling, smash the coarse veins on the larger leaves with the palm of your hand or a mallet. This will make them easier to roll.

Mince the onions and garlic. Melt the butter in a saucepan over medium heat and cook the onion and garlic until they are quite soft. Add the rice and cook for a minute or so, stirring to coat each grain with butter. Pour in the water and bring to a boil. Reduce the heat to medium-low, cover and simmer until the liquid is absorbed and the rice is tender, about 20 minutes.

Scrape the rice mixture into a large bowl and let cool for 5–10 minutes. Stir in the ground beef, salt and pepper, using a fork at first to distribute the rice, then switching to your hands to really bring things together.

Have a roasting pan and lid ready. Lay out 24 to 30 cabbage leaves.

Scoop about ¼ cup of the meat mixture onto the centre of each leaf, making adjustments to the quantity based on leaf size and condition. Using your hands, press the filling together. Fold the top and two sides of the leaf over the filling, then roll up from the bottom to form a neat bundle. Place seam side down into the roasting pan arranging half of the cabbage rolls in the first layer and nestling the remaining ones on top.

Pour in the tomato juice, making space between the rolls as needed so that it makes its way to the bottom. Top with the pineapple—the chunks are nicer looking, but crushed pineapple adds more flavour. Cover with the reserved leaves to prevent the rolls from scorching. Cover and bake at 350°F until tender, about 2 hours. I don't think cabbage rolls need to be served with anything—eat them on their own until you've had your fill. They keep for up to a week in the refrigerator.

Makes enough for 15 people.

Canada House Cazuela

CANADA HOUSE CAZUELA

A traditional South American slow-cooked dish with large chunks of meat and vegetables, cazuela is a one-pot meal that defines rustic elegance. Fernando's transcription of his recipe is one of my favourites. Written in Spanish, it's more of a shopping list with cooking times noted beside ingredient groupings. An illustration shows the type of zapallo (squash) needed, but there is no indication of method. The two times I was served it at Frei base it had been made with beef. The day I visited his kitchen he'd made the cazuela with chicken. I make mine from memory.

One beef eye of round (about 5 to 6 pounds) // 1 tablespoon coarse salt //
2 teaspoons cracked black pepper // vegetable oil

For the broth
2 onions // 6 cloves of garlic // 2 carrots // 1 sweet red pepper //
1 tablespoon each of dried thyme and oregano // 3 bay leaves // water

For the cazuela
6 medium-sized potatoes // 2 carrots // 1 medium butternut squash //
4 ears of corn // ½ pound green beans

Season the beef on all sides. Warm a large heavy-bottomed pot over medium-high heat and add enough oil to make a thin coating on the bottom. If you don't have a pot large enough, try a roasting pan and set it over two burners.

When the oil is hot put the meat in the pot and sear until the first side is well browned, about 3 minutes. Turn the beef onto another side and sear for several minutes. Continue in this manner until the entire roast is nicely crusted and the meat juices have caramelized along the bottom of the pan, about 15 minutes. Don't be afraid to go back over a side if you think you can get more colour from it.

While the meat is browning, prep the vegetables for the broth: leave the skin on the onions and cut them in two; smash the garlic, skin and all; peel and halve the carrots; seed and quarter the pepper. Arrange the vegetables around the roast as it finishes searing, then add the thyme, oregano and bay leaves. Pour in enough water to come halfway up the roast, cover partially and bring just to a boil.

Reduce the heat and gently simmer until tender, 3–4 hours. Turn the meat twice over the course of the cooking time and skim off fat as necessary. You'll also need to keep an eye on the level of the braising liquid and top it up with hot water so that the roast is always half immersed. When the meat is tender, remove the pot from the heat and set aside to cool. I like to cook the beef a day ahead and store it in the refrigerator immersed it in its broth until I'm ready to finish the cazuela. This makes it easier to remove any fat and renders the beef even more succulent.

To finish the cazuela

Skim any fat from the top of the pot containing the beef and lift the beef to a platter. Cover with plastic wrap and refrigerate until needed. Strain and discard the vegetables from the broth, then return the broth to the pot and set over medium-high heat.

While the broth is coming up to temperature prepare the vegetables for the cazuela: peel and halve the potatoes; halve and seed the squash, then cut each half into six wedges and remove the skin from each piece; peel and cut the carrots into oblique chunks; husk and cut the corn cobs into thirds; snap the tops from beans, but leave on the tails if they aren't too woody.

Add the potatoes to the pot, reduce the heat and gently simmer. After about 10 minutes add the carrots and continue cooking for another 5 minutes.

While the vegetables are simmering, carve the meat into 1 to 1½-inch thick slices—enough for one good-sized piece per serving. Return the meat to the pot, add the squash and simmer, covered, just until the beef is heated through and the vegetables are tender, about another 10 minutes. Time the addition of the beans and corn according to your preferences. I like mine crunchy so I add them 5 minutes before serving.

Serve cazuela in warmed shallow bowls. Arrange a slice of beef, a potato, a wedge of squash and a piece of corn in each bowl and tuck in the carrots and beans wherever you can. If your bowls are on the small side it will be a tight fit, but that's part of the charm. Ladle broth overtop and serve. When I have them on hand I like to tear fresh cilantro pieces or parsley into mine as I sit down at the table. My husband, who loves Asian cuisine, likes to stir in hot sauce—I imagine lime leaves would complete his culinary journey.

Makes enough for twelve people.

ALMOND BISCOTTI WITH ANISE SEED

The biscotti craze had just taken hold in Toronto when we left for Antarctica. Kevin, the chef aboard the Akademik Petrov, *indulged me one January day with a cookie tin full of his biscotti, and later shared the recipe. It calls for Pernod, an extravagance in any kitchen, so I often substitute almond extract. His original recipe made enough to serve an icebreaker's manifest.*

$\frac{1}{2}$ cup raw almonds // 1$\frac{1}{2}$ teaspoons anise seed // $\frac{2}{3}$ cup sugar //
$\frac{1}{3}$ cup butter at room temperature // 2 tablespoons vegetable oil //
2 eggs // $\frac{1}{8}$ teaspoon vanilla //
2 teaspoons Pernod (or 1$\frac{1}{2}$ teaspoons vanilla and $\frac{1}{2}$ teaspoon almond extract) //
$\frac{1}{2}$ teaspoon salt // 2 cups all-purpose flour // 2 teaspoons baking powder

Roast the raw almonds in a 350°F oven until you begin to smell their aroma, 8–10 minutes. While the almonds are roasting, crush the anise seed using a mortar and pestle or chop on a board with a knife. Allow the almonds to cool slightly. Chop them coarsely and set aside along with the anise seed.

Put the butter, sugar and oil in a medium bowl and beat together until light and fluffy. Beat in the eggs one at a time, followed by the salt, vanilla and Pernod (or vanilla and almond extract). Whisk together the flour and baking powder in a separate bowl, then stir them into the butter mixture to make a soft, sticky dough.

Preheat the oven to 325°F.

Turn out the dough onto a surface lightly dusted with flour. Coat your hands with flour and knead in the anise seed and chopped nuts. When the nuts are evenly distributed form the dough into a ball and cut it into four equal pieces. Roll each piece into a log about 12 inches long and place them about 3 inches part on lightly greased baking sheets. Gently press and pat along the length of each log to round the top, taper the sides and flatten the base.

Bake in the middle of the oven until pale gold and not quite firm to touch, 20–25 minutes. Remove from the oven and let cool on the baking sheet on a cooling rack for about 3 minutes before attempting to move. Use a long spatula to transfer the logs to a cutting board and slice diagonally into $\frac{3}{4}$ to 1-inch thick slices using a sharp chef's knife. Arrange the slices upright on the baking sheet about 1 inch apart. Return the pan to the oven and bake for another 20–25 minutes until they are golden, firm and dry.

Makes about 2 dozen cookies.

"Shackleton privately forced upon me his one breakfast biscuit,
and would have given me another tonight had I allowed him.
I do not suppose that anyone else in the world can thoroughly
realize how much generosity and sympathy was shown by this...
Thousands of pounds would not have bought that biscuit."

—Frank Wild. Wild was ill on the *Nimrod's* 1908 southern Journey.
Shackleton had put the four-man team on reduced rations.

Iceberg illuminated by flash bulb. Two expedition members stand on skis at
the iceberg's base. Ross Dependency, Herbert Ponting. June 4, 1911

POLARITY IS GOOD: DREAMING ANTARCTICA
Carol

Many people ask Wendy and me "How was it up there?" We have to reply, "Down there it was..." but it inspires us that people care about the Poles: the harbingers.

The Antarctica we experienced wasn't quite paradise. Nor was it only snowy and barren, as imagined. It was surprisingly green in places, multinational, busy and complex. It was ancient and a frontier simultaneously.

A vanguard Antarctic observer today, Elena Glasberg, says it is one of the most mediated locations anywhere.

That day on the Antarctic beach talking with people from five nations I knew what I wanted to do next: examine the interconnections between health, rights, environment and the law. Work outside more.

I've been concerned about Antarctica and the planet's health for years but recently my fear became visceral. I waited and waited for the snow to come for my urban cross-country skiing in Toronto. It barely came. According to NASA, "2012 was the ninth warmest of any year since 1880, continuing a long-term trend of rising global temperatures." Then in the winter 2014 we faced the inverse: a polar vortex freezing North America's Northeast. Extreme climate events are here.

We read daily about the sensitive polar regions of the Arctic and Antarctic warming rapidly and the devastating impact of ocean acidification. The human footprint in Antarctica is layered: local activity such as the introduction of invasive species and global activity of carbon dioxide emissions. There is still much to learn from the continent. What we do know is civilization faces water, food, land and human security crises. But we have survived past crises; we are the cultural remnants. There has been no war in Antarctica. This continent, for now, is reserved for dreaming and discovery.

Can eating together be good for the planet? We saw it in action. People told us stories.

There once was a base commander for the Argentineans in Antarctica. Their country had just lost the Falklands War. When British soldiers appeared at their station, the first thing the Argentinean commander did was tell the cook to turn on the stove.

WINTER IS COMING

MARCH 1–8, 1996

"Who could believe that those twenty barrels are filled
with little scraps of garbage—metal and glass and rusty bits,
and that a person picked them up bit by bit?"

MARCH 1, 1996

WT   5:00 a.m. arrival of new group—wish I hadn't stayed up so late baking. Too bushed to
hike with everyone. Shame we are on our way home so soon.

MARCH 2, 1996

Lena at about 11:30 a.m, "It's a good day for burning." Sun today, pretty well straight through,
same as yesterday. Last sauna.

Farewell cocktail party at Canada House with the Uruguayans, Chinese, Chileans and a
handful of Russians earlier tonight. The Chileans brought their kids—wonderful to see children
here and the way everyone was drawn to them, especially Walter the Uruguayan base commander singing the A-B-C song with them. Who knew it was the same tune in Spanish?
Exchange parting gifts. Great warmth. I was touched when Su Chong gave me a dumpling
rolling pin after everyone else had been given pins, badges or teacups from their gift shop.
"I know this is what you'd really like." Hope I can learn to make dumplings properly. Chileans
playing part of family men, quite distant now, Uruguayans flirty in the nicest way and
predictably last to leave. Happy to give everyone a braid of bread to take home. Last dinner
in Canada House and then on to a party at Sad Vlad and Ilya's. Just back from there now.

Guitar, chocolate and cognac the guys have brought in especially for me again. I know
it's the closest thing to wine they can find so I haven't got the heart to tell them I don't really
like it, and since they are so insistent that we eat when we drink, and drink when we eat,

that's what we do. Practised duet with Sad Vlad and then fell asleep at the top of the loft stairs. Escorted home by Hilltop Sasha and Ilya. No Sean; he left in a sort of a huff earlier. Met up with Tolya in his crazy raincoat. Not sure where he's off to. Must be the full moon.

MARCH 3, 1996

9:30 a.m. Last breakfast in Canada House. Rain in from the Drake today. Lots of wind; I love it when it's like this. Sitting on my spare bed, feeling the gusts through the wall.

Midnight. Blowy still. Can hardly keep my eyes open. Very busy day. Began tear-down and gruesome task of packing. Hope we can get everything loaded on the *Vavilov* without incident. Prepared food for tomorrow. Salads and sweets and lots of them as I want to spoil the guys. Lunch and dinner downstairs. Volodya made pancakes for lunch AND my favourite dinner too—all-in fried rice. Moving around kitchen with comfort. Feel part of these guys. A time of receiving gifts, tender glances. Volodya Driver telling me he loves me and will remember me. Up late packing and cleaning. Sorry to turn Sean away.

Roasted Pepper Goulash with Smoked Paprika
Caraway Noodles
Braised Cabbage
Red Cabbage Confit
Apple Crisp with Chantilly Cream

MARCH 4, 1996

2:00 a.m. Sad, lying in bed. The Drake is roaring, Dima says it's even darker during the day in winter than it is right now.

Rain in from the Drake today. Lots of wind. We've got some really early risers in this group—surprised to find people waiting for breakfast when I arrived, but managed to shoot from the hip and get everything done. Volodya Cook stood back and let me run. Sharing of kitchen fine today, as if we've been cooking together for years. Made bread for the last time— a huge batch. Kneaded dough into an elephant seal shape. Volodya and I made it yawn the way they do and then heave across the counter. Later I cut out bread men and bread women, one for each of us. Pork and borscht lesson with Volodya. Love the way he takes such care to sear each beet slice on the hot plate. Up the hill to Canada House for packing blitz.

Sean's got his hands full finishing the cleanup in the last three areas around camp so Sergey mobilized the masses to help me. The Russian guys are so bloody sound. Makes me angry to think of the people who complained the project wasn't much of a joint effort. We all have our roles. Joined by Lena, Dima, and Sad Vlad in the morning and afternoon. And like clockwork the truck arrives to take away the tables, fridge and stove. It will be hard to adjust to the outside world and all its static. Glad to have known this.

Visited Artigas this afternoon with the volunteers to say goodbye to our friends. How lucky we are to have been well received by all the bases. Sean pulled it together beautifully. And for the three of us to be accepted more like residents than tourists—brilliant!

Hat from Quique and the most beautiful letter I've ever received, especially with the mistranslation.

> March 1996
> Dear Wendy,
>
> You are a very nice person, without wickedness and you have good heart. When I am with you, I feel excellent. Because I enjoy when we are together, talking, dancing, etc. Each time I intercourse [sic] you more I know.
>
> People in general have the habit to put label to other people. Men aren't an exception, when they talk about women generally like sexual objects only. They compete to obtain this object. You aren't an exception for them. I disagree with them, and I feel sad because for me you are a lady and my friend. It's a mistake to include you. They murmur if you talk, you smile or dance with someone. You are a very especial person. They confuse how you are, because they aren't smart. They don't know you. You talk and you smile in the same shape with everybody. They don't know this.
>
> I would like to see you but I know is almost impossible because you get back to your country and I get back to my country. If you travel to Uruguay, visit me please. If I visit your country, if you wish, I will visit you.
>
> I think to be your pen pal.
>
> Three kisses
>
> Your friend, Quique
>
> P.S. is very difficult for me to write in English and more difficult than this to express my feelings.

Home late, just before 7:00 p.m. Scurry into the kitchen to find that the guys have already set the tables nicely for dinner. Quickly prep salads, bread, hors d'oeuvres.

Dinner a feast:
· Roast Leg of Pork
· Yummy Potatoes à la Volodya—Sean's favourite.
· Scads of salads: All-in Pasta Salad; Swiss Green Bean Salad; Russian Vinaigrette
· Honey Oatmeal Braids
· Kiwi fruit
· Frozen Chocolate Cream with Chocolate Sauce
· Biscotti
· Chocolate chip cookies

Many toasts. Sean's warm and sincere. Doc's genuine too. Sang "Milen'kij ty moj" with Sad Vlad and gave my toast. I love how bewildered they were when I spoke in full sentences, looking around the table, asking each other who'd been teaching me. I don't think anyone heard when I said, "You all have."—not that it matters. Merriment and laughing, especially at the sarong I made from the tattered Russian flag Radio Sasha gave me. Volodya Cook parading me all around the mess on his shoulders, I almost touched the sky. Guys helping me rescue the bread men and women from Bio Vlad. Man, he's moody. Not sure what he was on about collecting them from the tables, saying, "No one should eat bread in the human form." Guys distracted him while Sean, Lena and I had fun stashing the bread men in the library, the projection room, radio hut, Diesel and in the glove compartment of the PTS too. Who knows when they'll be found? Over to radio hut for cognac, tea and talk. To bed about 2:00 a.m. Really pushing it these last few days. First seizure I've had in a while.

Milen'kij ty moj
Verse 1

Woman sings	Миленький ты мой,	My darling
	Возьми меня с собой!	Take me with you
	Там, в краю далеком,	There in the faraway land
	Буду тебе женой.	I'll be your wife
Man sings	—Милая моя,	My sweetie
	Взял бы я тебя,	I would take you
	Но там, в краю далеком,	But there in the faraway land
	Есть у меня жена.	I already have a wife

Verse 2

Repeat first three lines from verse 1 and substitute last line with:

| Woman sings | Буду тебе сестрой. | I'll be your sister |
| Man sings | Есть у меня сестра. | I already have a sister |

Verse 3

Repeat first three lines from verse 1 and substitute last line with:

| Woman sings | Буду тебе чужой. | I'll be a stranger to you |
| Man sings | Чужая ты мне не нужна. | I don't need a stranger |

MARCH 5, 1996

Heavy heart. As we were finishing breakfast Radio and Hilltop Sashas, Dima, and Ilya told me how mariners believe if you whistle it will bring "nepogoda" (bad weather). I took it as a cue to push away from the table and head to the back stoop. They followed; we looked at each other and started whistling together. Sergey and Vadim joined in as they passed by— not that it worked. Calm water. *Vavilov* arrived early and departure went without a hitch, even with the dragging out of packing and goodbyes.

Dima, Sad Vlad and Hilltop Sasha to Canada House to help clean and pack. I don't think I thanked them—must do that as soon as I can by fax or telex. Gave my skis to Sad Vlad and treats to the others: rug, apron, magazines, vodka and candles. Knife to Volodya Cook and he gave me a chef's tunic, apron and hat. Found time to sneak away and go for a walk far up the beach to give my rocks and penguin parts back to Antarctica. If anyone saw me, I hope they understand.

Photographed the guys loading the PTS and watched it pull away from beach and fade into the mist. Lena saying, "It's so great; it's a great moment," as only she can. Later she told me one of the guys' reactions, "Who could believe that those twenty barrels are filled with little scraps of garbage—metal and glass and rusty bits, and that a person picked them up bit by bit?" If only all the volunteers could have seen this.

Goodbye hugs and kisses from all the guys gathered on the beach. Lovely of Roberto and Quique to come from Uruguay for the send-off. Passionate embrace from Volodya Driver, as if he meant it when he said he would take me with him, like in the song. And then onto the PTS. Lena, Sean and I straddling the packing crates. Volodya Cook running to the back stoop of the mess, saluting as we pass by. Guys on the beach, standing by the truck, waving. Not wanting to lose sight of them. Five flares shot into the air and Uruguayan helicopter flyby as we pulled out to sea.

Arrival at *Vavilov*. Glad to see familiar faces. Lena seasick. Toast to the guys at eleven as we promised we would.

Camp 17 Debris Collection (Feb. 29–Mar. 5)
Stoney Bay: $\frac{1}{2}$ barrel mixed waste & piping
Bellingshausen Areas 12, 15 & 5: 4 barrels mixed

> The people at Bellingshausen and the station itself should not be forgotten. Ours was a superficial start to tackling the basic problems of pollution and the tough work of containment of pollution and establishing better environmental conditions at Bellingshausen. I hope the work will continue. Only then will we have made a difference.
>
> —Volunteer Herb, Ontario

MARCH 6, 1996
Feeling emotionally fragile, clingy with Lena and Sean. What a trio we've become.

MARCH 7, 1996
Second full day at sea. Will switch back past the Cape and into port tomorrow. Feeling displaced. I still have one foot in Antarctica, as do Lena and Sean. Can't imagine what we'd feel like if the Drake gave us a rough go. We keep hearing stories from the volunteers about the *Petrov*'s crossing to the continent. They got that storm that prevented camp 16 from landing. Force 11, force 12 being a hurricane. I remember what it was like that night trying to fight my way down to the mess for dinner, the guys running to meet me to help me with my stuff and then being blown up the hill. Such a fine memory, I think I'll keep it with me.

Talked to Radio Sasha tonight, just after the *Vavilov* operator received a telegram with Women's Day wishes from the Diesel guys.

MARCH 8, 1996

Lovely day with Sean yesterday sunning on the deck, then dinner and report-writing. Amazing comfort I feel with this guy. Don't know when it started, but it's there now. He sees what I see, hears what I hear. Wished me Happy Women's Day at midnight, kneeling on the berth, porthole open, gazing out at the Beagle Channel, starlight and moonlight—Chile in silhouette. I guess it was Chile; we weren't sure.

Awoke in port. Dramatic step off the gangway onto continental soil. Spent morning in Ushuaia café-hopping and then a Women's Day lunch treat from Sean. Lena makes me laugh the way she defers. "Wendya you know what I'd like—you decide. You've been feeding me for three months." And later to me at Ushuaia airport, "I washed my hands with hot water and soap and there were paper towels—all these things I've dreamt of." Lena says to come to Saint Petersburg for the white nights around the solstice. Sean wants to go too. Funny about him—I have this feeling I could go anywhere in the world with him, more than with anyone I've ever known. But we both know that would be monumental.

Ran into Dave German just before we boarded. Good to see him and to hear that Outland New Forest has been asking if I'm looking to cook in the bush this spring. Don't know where my next dollars will come from, but I've got to give creativity a chance—trust that something will work out.

Sean and I locked together, watching a lightning storm in clouds below. A race through Buenos Aires airport. Then goodbye. Too hurried to be sad. Lena to me, "You and Sean have become so much a part of me, I sometimes find myself answering you in Russian." Must remember to tell Sean this.

Going solo—Solitude is an odd thing to search for in one of the most remote places on earth. And I laugh when I think how we hoped volunteers wouldn't miss what was around them while searching for something else. My life keeps reminding me we teach best what we most need to learn.

I loved the idea of venturing to the refuge on nights when all that nothingness swirled fiercely outside my window. Did I truly want to shut people out? Probably not. Free myself of cooking thoughts long enough to understand the poetry of KGI? Yes. But that is time and perspective, not solitude. It was pure folly to think I could hole up in a shack and divine the right texture or shade of blue to express Antarctica.

I think of a story Carol tells about a man she once travelled with in Poland. An avid diarist, he bowed out of excursions to write in his journal while the rest of the group wondered what he could be writing about. I think of all the colours I would have missed.

I've always been good at being alone. Years later, I realize Antarctica worked its way in and carved out a space in my heart as perfect as any refuge. Huddling close together down there, I wonder if the lesson Antarctica teaches best is not how to be alone, but how to be alone and together?

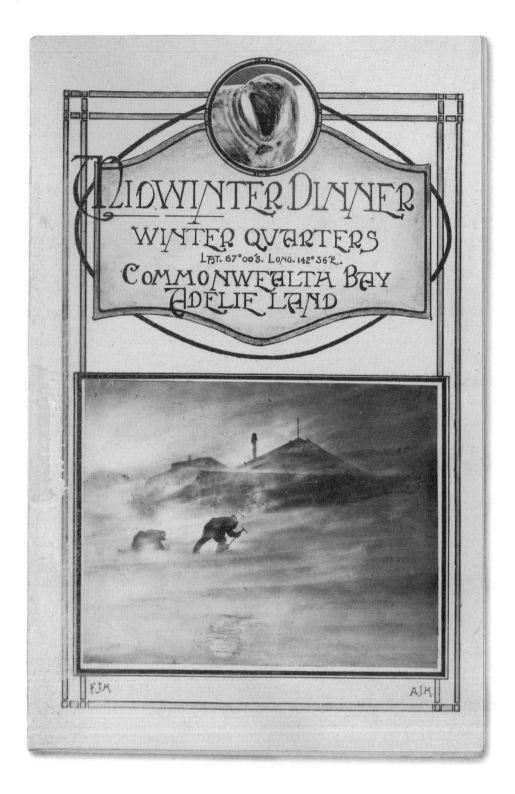

Menu prepared by Frank Hurley for Midwinter Day dinner, 22 June 1912,
Commonwealth Bay. Walter Hannam and Francis Bickerton prepared the meal.

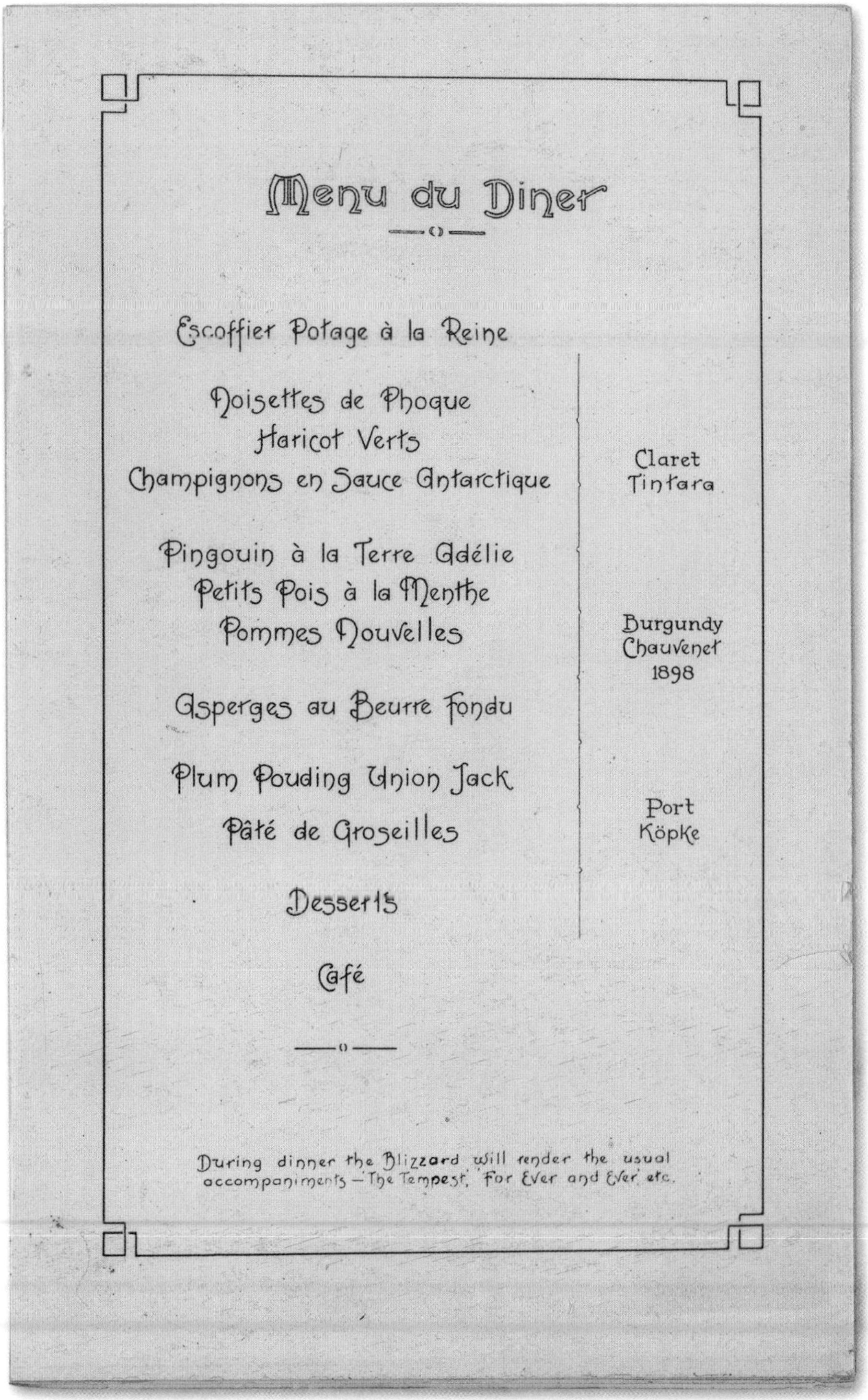

Menu du Diner
— () —

Escoffier Potage à la Reine

Noisettes de Phoque
Haricot Verts
Champignons en Sauce Antarctique

Claret
Tintara

Pingouin à la Terre Adélie
Petits Pois à la Menthe
Pommes Nouvelles

Burgundy
Chauvenet
1898

Asperges au Beurre Fondu

Plum Pouding Union Jack
Pâté de Groseilles

Port
Köpke

Desserts

Café

— () —

During dinner the Blizzard will render the usual
accompaniments — The Tempest, For Ever and Ever, etc.

ROASTED PEPPER GOULASH WITH SMOKED PAPRIKA

I was a mediocre stew-maker until I went to Antarctica, but with tender Argentinean beef at my disposal I turned myself around. Sure, it helped to be working with a budget that cancelled out my reticence to use wine, but it's the smoked paprika and sweetness of the peppers that bring the flavours together in this dish. I love that I don't have to use flour to thicken it and that when I made it for a garden gathering with dear friends, all the eight-year-olds at the table asked for seconds. Use hot smoked paprika if you prefer your goulash with a little more kick.

3 pounds stewing beef // olive oil // 3 medium onions // 6 cloves of garlic //
4 teaspoons caraway seeds // 4 teaspoons sweet smoked paprika // 1 tablespoon coarse salt //
cracked black pepper to taste // 1 can crushed tomatoes (28 ounces) // 4 cups water //
1 cup red wine // 4 sweet red peppers // 2 sweet yellow peppers

Cut the beef into 1½ inch chunks and pat dry. Chop the onions and garlic. Warm a large Dutch oven or heavy-bottomed pot over medium-high heat, and pour in just enough olive oil to make it slippery—if you use too much the meat won't caramelize and you'll miss out on all that rich colour. Brown the meat on all sides, then set aside. You'll want to do this in small batches so that each piece of meat comes in contact with the cooking surface with enough space between that the juices cook off. Otherwise you'll just steam the beef.

Add a drizzle of olive oil to the pot, followed by the onion, garlic and spices. Cook over low heat until soft. Return the meat to the pot, then stir in the crushed tomatoes, water and wine. Cover and simmer, stirring from time to time, until tender, about 1½ hours.

While the goulash is cooking put the washed and dried peppers on a baking sheet. Broil in the oven 3–4 inches below the heat source, turning them frequently, until the skins are black and blistered, about 12 minutes. Put the peppers into a bowl, cover it with a plate and let stand until they are cool enough to handle, about 20 minutes. Slip off the skins and remove the stems and seeds. Trim the ribs, cut the peppers into 1-inch squares and set aside until the final few minutes of cooking. When the meat is stewed to your liking—for me it's when the meat starts to fall apart—stir in the peppers and simmer uncovered until thoroughly warmed, about another 10 minutes.

Serve over egg noodles tossed with butter and toasted caraway seeds, topped with a dollop of crème fraîche or sour cream. For the gluten-free crowd, sauté 2 chopped onions and 1 shredded medium green cabbage in ½ cup of butter. Toss with caraway seeds. Cook until tender. Red cabbage confit is a delicious pairing as well—sometimes I serve both.

Makes enough for eight to ten people.

ROAST LEG OF PORK

Our first roast leg of pork was the perfect finale to a midsummer afternoon on the Uruguayan soccer pitch. I'm always a little baffled by praise for cooking a roast. This cut was so large that I had to ask Volodya Cook to trim off the hock to fit it in my oven. Aside from that all I did was stud it with garlic cloves, crumble herbs overtop and put it in the oven to cook slowly while I took my turn in goal. Canada House was standing room only for that meal.

Volodya and I prepared our final dinner at Bellingshausen in concert. I formed my honey oatmeal bread into bread men and women, much to the delight of whoever found them cooling and stood them upright on the baking sheet to dance. Volodya was on meat and potatoes. I watched as he cut all the meat from the bone of our last roast to stuff it with onions, garlic and herbs. It was delicious and I'm sure he used the bone to make stock and turned it into countless other meals. You can make an equally splendid roast without fussing as much.

1 pork leg roast (10 pounds with the rind on and hock off) // 1 to 2 heads of garlic // plenty of dried rosemary, thyme leaves and sage leaves // salt and ground pepper

Preheat the oven to 325°F. Pull the garlic cloves away from the bulb and peel each one, slicing any really large cloves into smaller pieces. Use the point of a knife to randomly cut into the rind of the leg as many 2-inch slits as you have pieces of garlic, and then insert a clove into each slit. (I take this same approach with any roast.) Put the pork into a large roasting pan and place in the oven. The total roasting time will be about 4–5 hours allowing 25–30 minutes per pound.

After three hours (about an hour before the roast has finished cooking) remove it from the oven. Use scissors to cut through the top of the rind. Slide a knife along the inside of the rind and pull it away from the roast as you go, leaving enough of the fat intact to make a nice crisp top. Sprinkle the herbs overtop, rubbing them between your fingers, and season with salt and pepper. If you're cooking for any crackling-lovers, tuck the rind back in beside the pork.

Return the roast to the oven and cook for about another 1–1½ hours until nicely browned and the juices run clear, or until a meat thermometer reads between 150°F and 160°F. Transfer the roast to a platter, cover loosely with foil and let rest for about 10–15 minutes before carving. The meat will continue to cook as it rests.

Makes enough for 20.

FROZEN CHOCOLATE CREAM

I guess cold desserts in cold places are counterintuitive, because it didn't occur to me to make ice cream until the last two camps. The first time out I simply froze my Chocolate Mousse. But for our going-away dinner I switched to this method with results more akin to the luxury of gelato. Perhaps if we'd been there a few more weeks I would have thought to make cones. Oh, hindsight.

7 ounces bittersweet chocolate // 3 cups whipping cream //
¼ cup sugar // pinch of salt

Have the chocolate finely chopped and ready in a bowl. Pour the cream into a heavy-bottomed pot over medium heat and warm it until it begins to steam and a few bubbles appear on the surface. Whisk in the sugar and salt and simmer until dissolved. Remove the cream from the heat immediately and pour it over the chocolate. Stir gently with a wire whisk to make sure that all the chocolate has come in contact with the hot cream; let it rest for a few minutes then stir slowly until smooth.

Pour the chocolate cream into a metal bowl or pan and cool to room temperature, stirring occasionally to keep a skin from forming. Cover with wax paper and chill in the refrigerator until cold (about 1½ hours), then move it to the freezer to harden. Let soften for about 5 minutes before serving.

Makes enough for six, or four true chocolate-lovers.

"June 22 was celebrated as it had been on *Endurance* with a feast, songs, and facetious sketches, all performed by the men in their sleeping bags. Like Shackleton, Wild took care to punctuate the monotonous existence with any excuse for an "occasion." Toasts were drunk to the King, the Returning Sun, and the Boss and Crew of the Caird with a new concoction consisting of Clark's 90 percent methylated spirit (a preservative for specimens), sugar, water and ginger..."

—Caroline Alexander, *The Endurance: Shackleton's Legendary Antarctic Expedition,* 1998

Midwinter Day Dinner June 22, 1911. Herbert Ponting, 1911

EXPECT VIOLET SWANS

Carol

———

Polish scientists discovered a black swan near their station. We commiserated over lunch at snowy Arctowski that it was unusual and sad: swans mate for life. Perhaps it wanted to go there. Scientists note ways 'vagrant' bird species reach Antarctica; austral winds push them off usual migration routes; they take ship-assisted journeys or pioneer new routes.

One scientist wondered if he could send the swan to Argentina on a ship. We do not know what happened to that swan.

The first record of the "black-necked" swan in Antarctica was in 1922, suspected to have wandered due to severe droughts. Another visited Antarctica in 1988. Black swans are native to Australia.

In Western society centuries ago, nothing was considered more impossible than a black swan. They were presumed non-existent until a Dutch explorer sighted one in Perth in 1697. This Antarctic experience taught me to expect even a violet swan.

Many of us do not live where we were born. Climate change and conflict will drive more human and species movement. Respect the land. Devote it to peace. Value science. Share. Protect wilderness and aesthetic values of nature. Ensure equitable, wise use of resources. Participate. Cooperate. These principles of a global commons for Antarctica (and everywhere) were alive on King George Island. Imperfect, but alive.

The Russian Antarctic Expedition staff did not choose us temporary workers, but they worked alongside the volunteers, opened their own impermanent home to us, and endorsed our humble cleanup efforts. It's worth making friends of neighbours because one never knows when one will need to catch a ship ride.

Until the end of that summer at Bellingshausen, everyone broke bread together. Wendy coached us to tear it with our hands.

Migration is not static: that's what the Russian scientists were observing in seals on the Fildes Peninsula. Massive elephant seals hauled past us into the ocean.

In bad weather volunteers cleared garbage from nearby the seal migration passages. It was only a tiny percentage of the iceberg of debris. We were happy the volunteers' work was recognized but more importantly we were honoured they joined the experiment.

Even with new and continued action, collaboration, respect for traditional knowledge, science and innovation we can anticipate surprises. The swans are coming, devastating and beautiful ones.

The party after five days and nights spent
in the open boats reach Elephant Isld.
They are partaking of the first hot food.

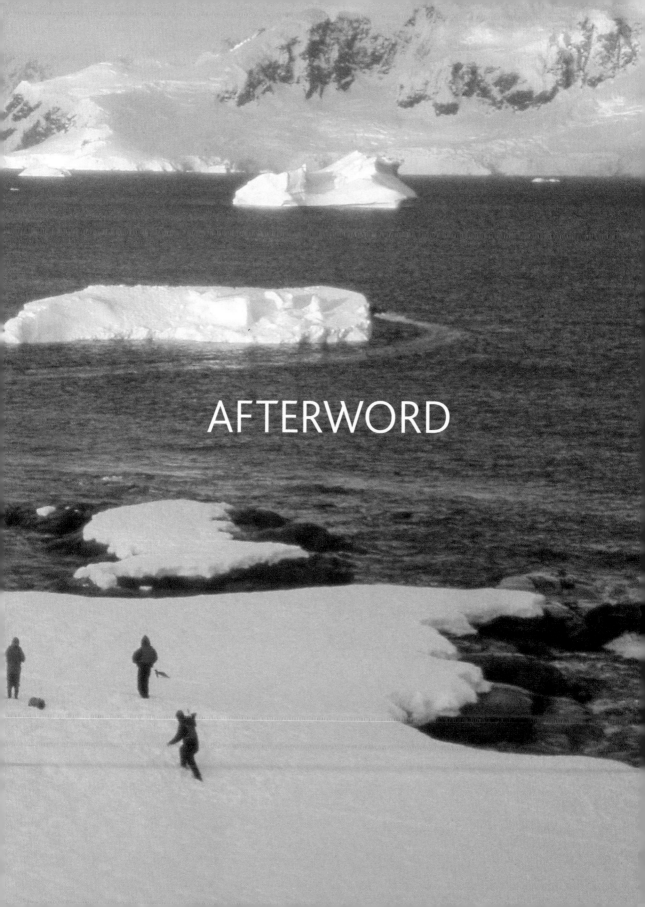

AFTERWORD

The final loadout: fuel pipe from Stoney Bay collected
during The Joint Ecological Project at Bellingshausen, March 5, 1996

August 16, 1998
Bellingshausen

Dear Wendy!

Sure you remember, a few years ago you were cooking in expedition in Antarctica. Maybe you would be interested to know an epilogue of this story.

Here writes Sasha, aerologist from Russian Bellingshausen station. Sorry for my scholar's English. It improved not a lot since that time.

I have had several months rest with my family after my 40th expedition (by the way we received on Bellingshausen your letter and photocollages, many thanks). In May 1997 I came to Bellingshausen by plane. Here are also several of our mutual familiars: Sasha Moroz our chief mechanique; Vladimir Kruzhkov, petroleum engineer; and last summer our station was visited again by Max Kirsling, German writer/tourist/philatelist. He also had good memories about that unique season when you and VIEW Foundation were here.

I found out from Vladimir you were at Bellingshausen once more, Antarctic summer 1997, and even made a barbeque for our winterers. It's a pity I'm late for it! Leonid our meteorologist told me you speak rather good Russian (Lena's lessons and the course you wrote us about are not lost for you). So interesting to know whether took place your vernissage and look at photos of your works, to know how you live.

Our Bellingshausen is planning to minimize. My greatest regret is aerology program operates last year. Jan. 1, 1999 it will close. For this reason I left for second wintering—maybe my last to Antarctica. I think due to this a project with volunteers was not continued. Many things remind me about that time, for example bread men you baked in the cinema cabin. And when we prepared Canada House for new inhabitants (Chilean builders), it was not yet weathered out a flavour either of spices or your perfumes.

A little about other persons of our crew. Vladimir, our sympatic and best guitarist, works in a photo studio in Saint Petersburg. Dima was going to go to King George Island for season, but did not. Sasha our radio operator from Ukraine wanted to go on this expedition, but he was refused because Arctic and Antarctic Research Institute's own radio operators are out of work. Vadim aerologist could not go due to health, same as Sergey, chief. They work in AARI. Lena our interpreter is working there also. She visited Bellingshausen on the airplane which brought us here in May 1997.

Be sure Wendy, that everybody on King George who remembered about you, did it with big warmth and sympathy. Your time spent here was not lost in vain.

—A. Kovalenko (Hilltop Sasha)

Letter from Lena Nikolaeva, St. Petersburg
May 24, 2000

My dear dear dear Wendy,
What a shame I did not write to you earlier. It may be due to my long stay in the safest place
on earth—Antarctica.

I am back home after a nearly five-month expedition.

What can I report to you? Antarctica is great and fascinating but it's not the fresh news.

Our stations are still behind in terms of environment protection and living conditions.

What is new? You would not recognize Bellingshausen. The buildings are newly painted
in red, all the old and shabby ones are demolished and removed, the old mess room (the one
filled with an ancient radio equipment) is repaired and newly furnished.

I celebrated New Years in Prydz Bay on board the *Akademik Fedorov*. No turkey this time,
but yes lots of Champagne. We started by midnight in the mess room with all the ship's crew
and expedition members present but in an hour or so split into small parties. I had no choice
but to join my inspection team and the Captain.

And now to the really big news. I mean that you are going to marry Cam. I am so very glad
that you are happy with him. Please write me more about your plan for the ceremony.

You are asking about our wedding customs. Though I married three times I don't know
much about the specifically Russian traditions.

At Russian wedding party usually at the beginning of it one of the guests should say
loudly that Champagne tastes bitter and this "criticism" becomes a signal to all the guests
to scream "bitter". In response the newlyweds should kiss each other to make it "sweet."
The kiss should be good enough to satisfy guests, otherwise they would keep screaming.

On my arrival to the office in April I found apart from yours, Christmas greetings from
John and Carol. It was very touching they remember me. I have not heard from Sean for
a long time but I do know that he had visited the Ukrainian station Vernadsky and met
Sasha, the station leader (do you remember Sasha, the Bellingshausen radio operator?
The one who drew funny cartoons).

Sasha writes to me from time to time. So, this world is very small and close.

Wendya, a lot to say and to ask you about but I have to finish now.

Hugs and kisses from St. Petersburg,
—Lena

REVERBERATIONS

Wendy

———

The lines between the end of one project and the beginning of another are often blurry. Perhaps it makes more sense to think of it as one big project with recurring themes? For me they are art, food, landscape and finding where stories dwell.

Even before I left for Antarctica I was considering what my response to the bottom of the world would be and how I might tell the story. Recipes were one way, I thought, but from my journals I can see I continued searching for a medium.

Soon after I arrived home I abandoned the idea of adding a chapter to the mixed media piece I'd been working on. *Dancing in a Northern Kitchen* (2003) was no place for Antarctic recipes and I knew there was more to tell even if I hadn't yet grasped the full scope.

The work that first emerged (and the vernissage Sasha asked about) was an installation that grew out of insights gained through journal-keeping. Best described as fragments that hover between abstraction and representation, inscribed with lace-like text, *Antarctic Chronicles* (1998) is an exploration of the role language plays in coding memory. Illuminated by candles, the collection is a meditation on the fragility of the environment and the fragility of memory.

Food has always been a means to an end, supporting my art. Since art college I'd also nurtured an idea to make an outdoor oven to explore connections between art, bread and dance. *Bread Breaking Boundarie*s (2000) came to fruition when I embraced the spirit of a Canada House dinner party and brought a community together through the simple act of baking bread.

In 2010, I finally found a place for the *Letters Never Sent* to Peter Gzowski in a project that examines the essence of letter writing. Part art installation, part performance and part travelling archive, *Voices at Hand* houses nearly 4,000 letters spanning 225 years.

My constant companion south of 60°, my journals were only intended for my eyes. It took me years to overcome my self-consciousness and open them. Our bookmaking journey helped me understand that the only way to serve the raw beauty of the experience and tell the story Carol and I most wanted to tell was to let memories closed in my heart have a life of their own. At the core we wrote this book to help our children understand that whether we are taking care of each other or taking care of the land, small gestures matter.

These days my cleanup projects are centred on my 100-year-old home, and picking up litter on walks to and from school with my son. His idea. In the midst of spring cleaning I rely on lessons we learned on King George Island—don't get overwhelmed, do a little at a time, and why not start with the basement. There is wisdom to be gleaned from sound housekeeping practices.

ANT/ARCTIC
Carol
SEPTEMBER 7, 2013

———

Antarctica is our memory.

—Laurence de la Ferrière, explorer and mountaineer. The first person and first woman to cross Antarctica (2000)

I'm flying over Baffin Bay in the Northern Atlantic Ocean. I'm en route to an Arctic health conference in Greenland. The midnight water below between these massive islands looks cold. Suddenly there is a bright white-blue dot in the ocean. It's a huge iceberg, floating from Greenland to Labrador, Canada where others might see it. The berg makes me smile.

I think of the first iceberg I saw, that aquamarine beauty floating one diamond clear Antarctic morning. I recall one night when I ran up the icy hill at Bellingshausen to see Vadim at the aerology hut. He spread the huge ivory balloon on a table and deftly fit the neck over the gas outlet nozzle. Vadim sent the balloon up, the razor wind carrying it into the clouds where the sun still refused to set. This was only one of millions of scientific inquiries to better understand our planet, our home and ourselves.

In the Antarctic I became more aware I live in a polar region. Climate has always shifted but glaciers melt faster than all predictions. We face health and wellbeing consequences, especially in lower-latitude and coastal areas from Greenland to Bangladesh.

Climate researcher Natuk Lund Olsen tells us that food is health, culture, identity and memory, "A Greenlander who has not eaten traditional Greenlandic food for a while experiences 'a kind of insatiable hunger for just this taste.'"

Inuit leaders urge for respect for their rights as stewards of the land, for negotiating power and for sustainable development. Spring, or *Upingarqsaaq* (Traveling to the Floe Edge) arrives earlier in Nunavut. Families fish by boat when previously they hunted from the ice floe.

We don't know yet who will most benefit from a summer ice-free Arctic and the resultant land and resource claims but development must be healthy and equitable. Northerners echoed Antarctic scientist's survival advice: collaborate, be alert for changing weather and don't put all your eggs in one basket.

When I think of Antarctica I recall recent images from a cleanup of abandoned materials at Distant Early Warning sites in the Arctic during the cold war. The rows of barrels filled with debris reminded me of the Bellingshausen garbage, collected by people who hadn't left the mess. We've run out of places for garbage.

In Antarctica, everything is stripped down. You have what you have and even less than that materially. It is only who you are and what you do that counts. Wendy jokes that her recipes constitute the 1,000 mile diet, but in fact her cooking at Bellingshausen reflected doing the best with what you have. Our expedition's hardships on King George Island were few but our opportunities to see humanity at its best were many—from the bottom, looking up.

RECIPE INDEX

IMAGE AND ARCHIVE CREDITS

Care has been taken to trace ownership of all archival material in this book. The publisher will gladly receive any information regarding errors or omissions.

Carol Devine, 14, 32 top, 42, 64–65, 74 bottom, 142 top, 240–241, 260–261

Sandy Nicholson, principal food photography, front cover endpaper, front cover endpaper (verso) and pages 6, 8–9, 16–17, 36–37, 78, 100–101, 106, 134–135, 138, 164, 198–199, 272, back cover endpaper (recto)

Wendy Trusler, pages 26–27, 32 bottom, 38, 46, 50, 72–73, 104, 128, 130, 142 bottom, 160–161, 166, 168, 170, 202, 204, 208, 210, 212, 262, back cover endpaper

The images by Frank Hurley appear by permission of Mitchell Library, State Library of New South Wales on page(s):
· 10, Hurley photographing under the bows of the "Endurance" / Call no. ON 26/2
· 24, The Bosun of the "Endurance", Vincent A.B. mending a net / Call no. ON 26/5
· 98, A radiant turret lit by the midsummer midnight sun / Call no. PXD 158/36
· 158, Shackleton (right) & Hurley...improvised cooking stove / Call no. PXA 715/18
· 196, Washing up after dinner at the winter quarters / Call no. Home and Away – 36823
· 248–249, Midwinter menu 22 June 1912 / in Charles Francis Laseron – Diaries Call no. MLMSS 385, Midwinter Dinner, Adelie Land / Frank Hurley Call no. Home and Away
· 258–259 The party after five days and nights spent in the open boats reach Elephant Island / Call no. PXA 715/20

The following images by Herb Ponting are licensed with permission of the Scott Polar Research Institute, University of Cambridge on page:
· 238, Flashlight Photograph of The Castle Berg. June 4, 1911 P2005/5/433
· 256, Midwinter Day Dinner at Winterquarters Hut. June 22nd 1911 P2005/5/447

The image on page 62 from the Charcot expedition, appears with permission of Anne Marie Vallin-Charcot and thanks to Serge Kahn.

The image of Carol Devine and Wendy Trusler on page 74 was taken by Lena Nikolaeva at Bellingshausen in 1995.

The image on page 132 of Edith Jackie Ronne and quotation from her journal appear by permission of Karen Ronne Tupek.

The excerpt from "Trespassing on Eternity" on page 129 appears by permission of Bob Payne and *Condé Nast Traveler.*

ACKNOWLEDGEMENTS

We could not have done the project without the adventurous spirit, dedication and hard work of the VIEW Foundation volunteers and the support of Bellingshausen's 1995–1996 staff, the Russian Antarctic Expedition and MEI. Merci beaucoup.

Special thanks to Sandy Nicholson for his steadfast support, tenacity and exquisite photography.

Thank you Patrick Shaw, Mickey and Marjorie Devine, Marguerite Hoffman and Tim Gale for the generous support that made this book possible.

We would also like to extend our sincere appreciation to: David Young, our editor, for his insight, clarity and championing our project. The Office of Gilbert Li, designers Isabel Foo and Gilbert Li, for their brilliance, vision and endurance; Donna Bartolini our recipe editor, for sharing her cooking wisdom and encouragement and Nancy Payne for her copy editing and proof reading.

We are grateful to Charlotte Sheedy and Mackenzie Brady of Charlotte Sheedy Literary Agency for believing in and representing our book.

Thank you Elizabeth Viscott Sullivan, Executive Editor at HarperCollins Publishers. We are thrilled about this new part of the journey.

Special recognition to the following individuals and institutions: Lena Nikolaeva, Dr. Valery Lukin, Victor Pomelov and Arctic and Antarctic Research Institute (AARI); R.K Headland, Heather Lane, Naomi Boneham, Lucy Martin and Dr. Paul Berkman of The Scott Polar Research Institute; Kevin Leaman, State Library of New South Wales; Mitchell Library; Lynn Lay, Byrd Polar Research Centre, Ohio State University; Soojin Creative; Stephanie Nolen; Karen Connelly; Anne-Marie Vallin-Charcot; National Library of Australia; In the Footsteps of Douglas Mawson Exhibition, South Australia Museum; Cris de Boos, Erskine Press; Scientific Committee on Antarctic Research (SCAR); International Polar Year 2012; Association of Polar Early Career Scientists, and Polar Professionals.

Thank you to Lita Albuquerque, Sara Angel, Sam Blyth, Sarah Brohman, Barbara Chambers, Barbara Chisholm and Thomas Miller, Bryce Conacher, John Croom, Jean de Pomereu, Maj de Poorter, Michael Devine, Carmen Dunjko, Mark Epstein, Steve Falk, Molly Finlay, Kate Gammal, Ruth Gangbar, Susan Heinrich, Paul Jerinkitsch, Andrea Juan, David Kennedy-Cutler, Serge Kahn, Doug Laxdal, David Lightfoot, Cornelia Lüdecke, Doug McIntyre, Melanie Ostry, Bob Payne, Barb and David Russell, John Spletts, Sean Steven, Karen Tupek, Frank Viva and Tammy Yiu (OSC Cross) for their valued contributions, enthusiasm and in some cases taste testing. Thank you also to the many others, too numerous to mention, who shared ideas, loaned props or broke bread with us at post shoot feasts and more.

And finally, thank you to Mickey, Marge, Sasha and Veronica Devine, Meg and Garth Nicholson, Carol and Bill Trusler and Cameron and Finley Taylor for your faith and your love.

BIBLIOGRAPHY

"Antarctic Climate Change and the Environment – 2011 Update, SCAR, XXXIV," Antarctic Treaty Consultative Meeting, Buenos Aires, June 20–July 1, 2011 (www.scar.org).

Alexander, Caroline, *The Endurance: Shackleton's Legendary Antarctic Expedition*, New York: Alfred A. Knopf, 1998.

Behrendt, John C. "Contrasts in Antarctica Since IGY: Without and With Women," Institute of Arctic and Alpine Research, University of Colorado, American Polar Society Symposium, Women's Roles in Polar Regions, Past, Present and Future, October 9–10, 2003.

Berkman, Paul Arthur, *Science into Policy: Global Lessons from Antarctica*, San Diego: Academic Press, 2002.

Bloom, Lisa et al. eds. *Gender on Ice*, The Scholar & Feminist Online webjournal. Barnard Center for Research on Women, 2008. http://sfonline.barnard.edu/ice/bloom_01.htm (accessed January 20, 2012).

British Antarctic Survey, "The Antarctic Treaty Explained," http://www.antarctica.ac.uk/about_antarctica/geopolitical/treaty/explained.php (accessed May 27, 2012).

Challenger, Melanie, *On Extinction: How we became estranged from nature,* London: Granta, 2009.

Charcot, Jean-Baptiste, *Towards the South Pole Aboard the Français; The First French Expedition to the Antarctic 1903–1905*, Bluntisham: Bluntisham Books; Norwich: Erskine Press, 2004.

Cherry-Garrard, Apsley, *The Worst Journey in The World*, London: Penguin Classics, 2006 (first published 1922).

Environmental Services Association of Alberta (ESAA). "Antarctica Cleanup, The Story of Bellingshausen Station," Prepared for Robert Swan and 2041, 2011. http://www.2041.com (accessed April 29).

Forbes, L. M. ed. *Making Meals With Seals: Recipes from the Table of the British Graham Land Expedition 1934–37*, Scott Polar Research Institute, 2011. *The Polar Record*, Vol 9 Nos 58–63, 1958–59, Cambridge: Scott Polar Research Institute, 1960.

Glasberg, Elena, *Antarctica as Cultural Critique: The Gendered Politics of Scientific Exploration and Climate Change*, Hampshire: Palgrave Macmillan, 2012. http://www.palgrave.com/products/title.aspx?pid=523946 (accessed April 26, 2012).

Glasberg, Elena. "Refusing History at the End of the Earth: Ursula Le Guin's 'Sur' and the 2000–1 Women's Antarctica Crossing," *Tulsa Studies in Women's Literature*, 21, 1. Spring 2002.

Hansson, Heidi and Cathrine Norberg, eds. "Cold Matters: Cultural Perceptions of Snow, Ice and Cold." *Northern Studies Monograph* no. 1, Umeå University, 2009.

Hurley, J. Frank, *Diary of Imperial Trans-Antarctic Expedition 1914–1917, 5 November 1914–31 December 1915*. National Library of Australia.

Khaleefa, Abdulwahab. "An Analysis of the Short Story 'Sur', Report of the Yelcho Expedition to the Antarctic, 1909–1910," http://www.ursulakleguin.com/Sur-Analysis.html (accessed April 15, 2012).

Korczak-Abshire, Małgorzata et al. "First documented record of barn swallow (Hirundo rustica) in the Antarctic," Zakład Biologii Antarktyki, Polska Akademia Nuak, Warszawa, Poland. *Polish Polar Research*, vol 32. no. 4, 2011.

Le Guin, Ursula K. "Sur." *The New Yorker,* February 1, 1982.

Marsching, Jane, and Andrea Polli, eds. *Far Field: Digital Culture, Climate Change, and the Poles*, New York: Intellect, 2011.

Mawson, Douglas, *Home of the Blizzard, The Story of the Australasian Antarctic Expedition, 1911–1914*, Adelaide: Wakefield Press Pty, Limited, 1930.

May, John, *The Greenpeace Book of Antarctica: A new view of the seventh continent*, Toronto: MacMillan of Canada, 1988.

McEwin, Emma, *An Antarctic Affair: A Story of Love and Survival as Told by the Great Granddaughter of Douglas and Paquita Mawson*. Millswood: East Street Publications, 2008.

Miller, Paul D, *The Book of Ice*, Brooklyn: Mark Batty Publisher, 2011.

Mills, Leif, *Frank Wild*. Yorkshire: Caedmon of Whitby, 1999.

Montalti D., Orgeira J. L. and di Martino S. "New records of vagrant birds in the South Atlantic and in the Antarctic." *Polish Polar Research,* vol 20: 347–354, 1999.

Nansen, Fridtjof. Introduction to *The South Pole An Account of the Norwegian Antarctic Expedition in the Fram, 1910–1912,* Roald Amundsen, London: John Murry, 1912.

Norgaard, Kari Marie, *Living in Denial: Climate Change, Emotions and Everyday Life*, Boston: MIT Press, 2011.

Payne, Bob. "Trespassing on eternity." *Condé Nast Traveler,* January 1997.

Rawlinson, Mark. "'Waste Dominion', 'White Warfare', and Antarctic Modernism." *Tate Papers,* Issue 14, Autumn 2010.

Riffenburgh, Beau, *Racing with Death: Douglas Mawson–Antarctic Explorer*, London: Bloomsbury, 2008.

Roberts, Lisa. "Antarctic Animation: Gestures and lines describe a changing environment." Doctor of Philosophy diss., College of Fine Arts, University of New South Wales, Oct 2010, http://www.antarcticanimation.com (accessed Jan 14, 2012).

Ronne, Edith M. "Jackie", *Antarctica's First Lady: Memoirs of the First American Woman to Set Foot on the Antarctic Continent and Winter-Over,* Stony Brook: Celebrity Profiles Publishing Co., 2004.

Rubin, Jeff. "Train Oil and Snotters Eating Antarctic Wild Foods," *Gastronomica,* 2003.

Scott, Robert F., *Scott's Last Expedition, Volumes 1* and *2.* http://www.gutenberg.org/files/11579/11579-8.txt

Smith, Roff, *Life on the Ice: No One Goes to Antarctica Alone,* Washington, D.C.: National Geographic Society, 2005.

Shackleton, Ernest. *The Heart of the Antarctic: The Story of the British Antarctic Expedition 1907–1909,* London: William Heinemann, 1909. http://archive.org/details/heartofantarctic02shac (accessed May 2, 2012).

Swan, Robert and Gil Reavill, *Antarctica 2041: My Quest to Save the Earth's Last Wilderness,* New York: Crown Publishing Group, 2009.

Spufford, Francis, ed. *The Antarctic,* volume 2 in *The Ends of the Earth: An Anthology of the Finest Writing on the Arctic and the Antarctic.* London: Granta Books, 2007.

Wheeler, Quentin, "New to nature No 70: Solanum baretiae." *The Guardian,* April 8, 2012, *The Observer,* http://www.guardian.co.uk/science/2012/apr/08/jeanne-baret-tribute-solanum-baretiae (accessed Jun 5, 2012).

Yusoff, Kathryn. "Visualising Antarctica as a place in time; from the geological sublime to 'real time'." The Surrey Institute of Art & Design/Royal Holloway, University of London, Visual Knowledges Conference, University of Edinburgh, September 17–20 2003.

ABOUT THE AUTHORS

Carol Devine is a researcher, writer, activist, and humanitarian professional. She is strategic advisor to the Museum of AIDS that aims to open in 2017 in South Africa. She has also worked for Médecins Sans Frontières (MSF) in Rwanda, Southern Sudan, and East Timor and was MSF Canada's Access to Essential Medicines Campaigner, fighting for accessible, effective, and affordable medicines for developing countries. Carol created the Antarctic civilian clean-up expedition on which this book is based. She lives in Toronto.

Wendy Trusler is an interdisciplinary artist, designer, writer, and food stylist. The expedition cook in Antarctica, she is the author and stylist of the forty-two recipes in this book. For more than twenty years she has balanced her work as a cook and artist, cooking and catering, styling food for film and television, and developing her art practice driven by ideas related to ecology, continuity, and regeneration. She lives with her husband, son, and Shackleton the cat in Peterborough, Canada.